www.KnowledgePublications.com

SAVE HARMLESS AGREEMENT

Because use of the information, instructions and materials discussed and shown in this book, document, electronic publication or other form of media is beyond our control, the purchaser or user agrees, without reservation to save Knowledge Publications Corporation, its agents, distributors, resellers, consultants, promoters, advisers or employees harmless from any and all claims, demands, actions, debts, liabilities, judgments, costs and attorney's fees arising out of, claimed on account of, or in any manner predicated upon loss of or damage to property, and injuries to or the death of any and all persons whatsoever, occurring in connection with or in any way incidental to or arising out of the purchase, sale, viewing, transfer of information and/or use of any and all property or information or in any manner caused or contributed to by the purchaser or the user or the viewer, their agents, servants, pets or employees, while in, upon, or about the sale or viewing or transfer of knowledge or use site on which the property is sold, offered for sale, or where the property or materials are used or viewed, discussed, communicated, or while going to or departing from such areas.

Laboratory work, scientific experiment, working with hydrogen, high temperatures, combustion gases as well as general chemistry with acids, bases and reactions and/or pressure vessels can be EXTREMELY DANGEROUS to use and possess or to be in the general vicinity of. To experiment with such methods and materials should be done ONLY by qualified and knowledgeable persons well versed and equipped to fabricate, handle, use and or store such materials. Inexperienced persons should first enlist the help of an experienced chemist, scientist or engineer before any activity thereof with such chemicals, methods and knowledge discussed in this media and other material distributed by KnowledgePublications Corporation or its agents. Be sure you know the laws, regulations and codes, local, county, state and federal, regarding the experimentation, construction and or use and storage of any equipment and or chemicals BEFORE you start. Safety must be practiced at all times. Users accept full responsibility and any and all liabilities associated in any way with the purchase and or use and viewing and communications of knowledge, information, methods and materials in this media.

Hydrogen Generator Gas for Vehicles and Engines: Volume 5

PRODUCER GAS FOR MOTOR VEHICLES

By

JOHN D. CASH

and

MARTIN G. CASH

ANGUS AND ROBERTSON LTD
SYDNEY :: LONDON
1940

ISBN: 978-1-60322-035-4

We are happy to have been able to bring such an excellent work of text back to life and to once again make it available

Set up, printed and bound in Australia by Halstead Press Pty Limited, 9-19 Nickson Street, Sydney 1940

Registered in Australia for transmission through the post as a book

INTRODUCTION

Australia, though abundantly blessed with material resources, with pastoral, agricultural and mineral wealth, has so far been denied one of the most necessary resources—natural oilfields.

In the present time of national emergency the question "What can we do to remedy this natural deficiency?" becomes of vital importance.

Motor fuel is a modern necessity. Without it transport is paralysed, and without transport industry cannot function. Under present conditions it is problematical whether our cities could be fed, or our rural industries maintained without the use of motor vehicles.

One does not need to be an alarmist in order to visualize a set of circumstances in which the supply of motor fuel to Australia from overseas may become impossible. Lack of shipping space; lack of credits, or the necessity of conserving our international balances; the entry of a strong Pacific maritime Power on the opposite side in the international conflict; any, or indeed all, of these emergencies may conceivably arise, and cut off our supplies of imported fuel.

Have we then a natural substitute for our imported motor fuel? The answer is yes, at least one (probably more), and the purpose of this book is to set forth the possibilities of this substitute—producer gas—and the methods by which it can be brought into use in a national emergency.

It may be asked why producer gas has not come into general use and displaced imported fuels in times of peace. The reasons are many, but four predominate: (i) Higher first cost, (ii) cumbersome equipment, (iii) loss of power as compared to petrol, and (iv) lack of flexibility.

That producer gas running costs are lower than petrol costs is undeniable. But the easy starting of the petrol engine, its flexibility of control, the smaller amount of space required for the carriage of fuel, as well as the network of pumps whereby petrol supplies are ensured at all times and all places (in other words the selling technique of the oil companies), has kept imported motor fuels in the foreground, and producer gas has remained in the background—a scientific curiosity only.

In the year 1860 Lenoir invented the first successful gas engine, which operated on coal gas. This gas was expensive and required large equipment for its manufacture. The natural outcome of this was the introduction of a system whereby large quantities of gas could be made quickly, with comparatively small equipment and from low-priced fuels such as coke, charcoal, wood etc. The gas producer satisfied all of these points and so was slowly introduced.

Among the earliest recorded gas producers were those of Birchof (1839) and the Siemens brothers (1857). It is probable that both these plants were of the pressure type. The first suction gas producer was invented by Dr Jacques Arbos of Barcelona in 1862.

The history of the motor car itself extends back little farther than the beginning of this century, and the history of the gas producer as applied to the motor vehicle almost as far.

Between the years 1901-1903 an Englishman named Parker drove first a 2½ h.p. and later a 25 h.p. motor car over a distance of 1000 miles with results not wholly unsuccessful.

When the 1914-1918 war broke out the mobile gas producer had not reached the stage where it was even considered as a substitute for petrol. Consequently the war retarded its advancement and little more was heard of it until the high price of petrol in England about 1922 gave it a new lease of life.

With the onset of the economic depression in Australia in 1929 and the sudden rise in the price of petrol, due to a combination of increases in duty and tax and to the adversely altered exchange value of Australian currency, a new interest in the gas producer for motor vehicles was aroused, more particularly in South Australia and Western Australia.

By 1930-1931 a number of plants had been fitted to tractors and trucks and even a few cars. These plants were experimental and usually very cumbersome.

With an improvement in economic conditions and a fall in the retail price of petrol in the years 1933-1935 interest in the gas producer waned to a considerable extent. This comparative lack of interest continued until the outbreak of the present war when producer gas began to hold unparalleled interest for all owners of petrol-driven vehicles, and indeed for the community in general.

The modern gas producer is made up of three distinct elements, the efficient functioning of each being the deciding factor in the results achieved by the plant as a whole. These elements are:

1. The generator in which the fuel is burnt to produce the gas ultimately used in the engine. This is usually a cylindrical or rectangular metal container holding the fuel and into which air is admitted by means of either a grate or tuyere.

2. A scrubbing or cleaning element which removes from the gas any of the impurities it may contain. The means of achieving this result are either dry or liquid cleaners, or in some cases a filter through which the gas is drawn.

3. A cooler or radiator. As the gas is generated from the fuel by means of a fire, the temperature of which is very high, it will have a correspondingly high temperature which results in a loss of engine power if it is not first cooled down to a temperature of about 130° F. This is done in the cooler, or radiator as it is often called, by passing the gas through a number of pipes which are air-cooled.

In addition to the units just mentioned there must be some apparatus in which the producer gas and air mix to form a combustible mixture.

That a gas producer is an inconvenient, bulky, and in many cases, unsightly, piece of apparatus must be admitted. But some of this prejudice must vanish when we consider the important task that it is performing.

Nature herself took untold thousands of years to produce petroleum, and it is indeed a blessing that man, with the aid of a very simple apparatus and some wood, can produce an efficient substitute. Giant hydrogenation plants capable of producing a satisfactory substitute fuel from coal have been built at enormous cost; but the producer gas plant is the only plant that can produce a satisfactory substitute for petrol and be built within the means of a private individual.

It is hoped that this book may add something to the general knowledge of this subject, and assist in making the motoring public give more thought to producer gas. It will make available to potential purchasers of producer gas plants information which will enable them to choose wisely, and protect them from the extravagant claims of over-enthusiastic salesmen. Engineers, garagemen, students, and experimenters will find the book helpful and thought-provoking.

Finally, if the information and suggestions contained herein contribute even in small measure to the solving of Australia's war-time transport problem, the aim of the book will have been achieved.

JOHN D. CASH.
MARTIN G. CASH.

Kew, Victoria.
May, 1940.

CONTENTS

CHAPTER I

HEAT POWER AND ENERGY

SOME 60,000 years have passed since man discovered that fire could warm him and provide a barrier between him and the elements that sought to destroy him.

It was not, however, till 250 years ago that he seriously considered harnessing the power of fire for the production of mechanical energy. This he did by converting the energy contained in fuels such as coal and oil into heat and then into mechanical energy. The wheels of our present civilization have come to rely chiefly on these two fuels as sources of power, and thus a knowledge of heat, which is used to make this power available, is essential for a thorough understanding of the action of either an engine or, in this particular case, a gas producer.

This brief description of the principles of heat and energy is inserted to avoid an undue amount of reference to other works.

That mode of energy which we distinguish as heat is now regarded by physicists as a vibratory molecular motion taking place within the substance which is considered to be hot. Fortunately, to acquire a knowledge of the application of heat it is not necessary to have an understanding of the nature of heat itself.

TEMPERATURE

Temperature is a statement of the degree of hotness or coldness of a body compared with some other standard.

B

The measurement of temperature is effected by means of an instrument known as a thermometer which makes use of the fact that nearly all substances expand on being heated and contract on being cooled.

Two known states of temperature—namely the freezing and boiling points of water—have been chosen to provide a constant temperature difference, and various scales of temperature have been devised by taking this difference and dividing it into a number of equal parts called degrees. Only two of these scales however concern us now.

In the Fahrenheit (F.) scale the freezing point is marked 32° and the boiling point 212°; the interval between these points being divided into 180 equal parts, or degrees. Zero on this scale will thus be 32° below freezing point.

The Centigrade, or metric, scale has the freezing point marked 0° and the boiling point 100°, the interval being divided into 100 degrees. Engineers for the most part use the Fahrenheit scale, while physicists and chemists use the Centigrade scale.

Units of Heat

Various units of heat have been devised by considering a unit to be that quantity of heat required to effect a unit change in the temperature of a unit mass of water.

In the British system the British Thermal Unit is the unit employed, and is that quantity of heat required to effect a change of temperature of 1° Fahrenheit in 1 lb. of water. This unit is called the B.Th.U.

In the metric system the gramme-calorie is the unit of heat and is that quantity of heat which is required to effect a 1° Centigrade change in the temperature of a 1 gramme mass of water. More frequently a unit of heat 1000 times as great as the gramme-calorie is used, this being known as a calorie. It is important to note that temperature and heat are not the same thing—temperature being the degree of hotness or coldness of a body, while heat is the agent which produces the hotness.

Heat and Mechanical Energy

Both heat and mechanical energy are forms of energy that can be changed one into the other. Thus a certain amount of mechanical energy can be made to produce a certain amount of heat energy, and conversely a certain amount of heat energy can be made to produce a certain amount of mechanical energy.

Joule's Mechanical Equivalent

Joule discovered that the energy possessed by a 1 lb. weight falling through a distance of 778 feet (mechanical energy) can be made to raise a 1 lb. mass of water through 1° F.—that is, it is equivalent to 1 B.Th.U.

This relationship is referred to as Joule's mechanical equivalent of heat, since it expresses the amount of mechanical energy equivalent to a unit of heat energy.

Work and Power

Whenever a body is moved against the resistance of some force, work is said to be done. The amount of work is measured by the product of the force and the distance the body is moved against the force.

In engineering circles the unit of work is taken as the work required to raise a mass of 1 lb. vertically through a distance of 1 foot. This unit is called the foot-pound of work.

It will be noted that in the definition of work no mention was made of time. When time is considered we have power, which is the rate of doing work.

Horse-power

When James Watt began to sell his own steam engines he had no means of telling his prospective customers what power they could expect from the engines, and he therefore invented the term "horse-power" which was supposed to be equal to the power that could be exerted by a horse for a whole day. Either the horses in Watt's time were very much more powerful than those of to-day, or Watt's knowledge of horses was very scanty, for it is an exceptional beast that will exert a "horse-power" for a whole day.

Horse-power as expressed in the engineering sense is 33,000 foot-pounds of work done per minute, so that an engine capable of giving 10 horse-power would be able to lift 330,000 lb. through a distance of 1 foot or alternatively 1 lb. through a distance of 330,000 feet in 1 minute.

One horse-power for one hour is called "one horse-power hour."

$$1 \text{ h.p. hr.} = \frac{33{,}000 \times 60}{778} = 2560 \text{ heat units}$$

One horse-power hour = 2560 B.Th.U. at 100 per cent efficiency. But when using producer gas there is firstly the loss in the generator and cooler, and secondly the loss in the engine, which is only 25 to 27 per cent efficient, so that in all about 10,000 B.Th.U. are required to produce 1 horse-power hour.

CHAPTER II

THEORY OF PRODUCER GAS

THE chemistry of producer gas is an involved study. There are problems which may well tax the knowledge and ingenuity of even the most advanced industrial chemist, or research worker. However, by keeping to a few commonsense rules, those who wish to build a gas producer can do so and get good results without any knowledge of chemistry, in much the same way as one may light a fire and cook a good dinner without a knowledge of the chemical actions taking place in the fire or the vastly more complicated chemical changes taking place in the dinner.

The essential practical requirements (each of chemical significance) for a producer gas plant are as follows:

1. Air is drawn through a mass of red-hot or glowing charcoal contained in a closed firebox called the generator, where it is changed by chemical action to a poisonous and explosive gas.

2. The glowing charcoal must be sufficiently compact, so that air cannot pass through without making contact with the hot charcoal. Fortunately the vibration of a moving vehicle usually shakes the upper layers of charcoal down and prevents the "caving" or "arching" which might otherwise allow air to pass through the generator unchanged.

3. There must be a surplus of glowing charcoal so that the air must pass through not less than 6 inches of it or the

action may be incomplete, and a gas which is little better than smoke may be produced. If the generator is large enough this takes care of itself, as the fire soon spreads enough for complete action. Trouble from incomplete conversion of air to gas will arise if an attempt is made to use the generator until all the fuel is burnt away. Explosions in the scrubbers may also result from trying to use the last of the fuel in the generator for gas production.

4. Good results are obtained without the use of water in the generator, but if it is desired to generate water gas a small quantity of water may be admitted, preferably to the hottest part of the fire, either as steam or as a fine jet of water. For vehicle use when the load is intermittent, some means should be provided by which the water jet is turned off under light loads, otherwise the fire may be dimmed or even quenched. When the fire is hot, water may be admitted to the extent of one drop per second for each 2 h.p. that the engine is developing at the time.

5. Air must be mixed with the gas in order to produce an explosive mixture, and is usually admitted at the induction pipe or inlet manifold. The controls should be capable of giving an excess of air at full loads.

Each of the foregoing paragraphs contains a practical instruction with a chemical background which, if carried out, will give the required chemical results. No more than this is required for the ordinary operator, but for the designer or the experimenter who is trying to improve the gas producer for vehicles, a section on the chemistry of the subject is provided.

Chemical Change

A chemical change takes place as a result of forces inherent in nature. Man can only arrange the set of conditions which experience has shown to be necessary for a certain chemical action. The action then takes place spontaneously. Substances which unite readily when mixed are said to have chemical affinity. Substances which do not unite under any conditions, or can be made to do so only with difficulty, are said to lack chemical affinity. Even substances having chemical affinity cannot unite unless the physical conditions are favourable. These conditions are:

1. The substances must be in close proximity. Solids may be brought into close proximity by grinding, then mixing. Solids may be even more intimately mixed if each can be dissolved in the same solvent. Gases readily intermix by diffusion so that each particle of the one has an equal chance to encounter a particle of the other, thus giving the best possible condition for chemical action. Diffusion is rather a slow process and when very rapid chemical action is required, as in an engine cylinder, mechanical mixing or turbulence is desirable in addition.

2. The speed of chemical action is also affected by the concentration; that is, by the weight of each substance contained in a given volume. In the case of gases this varies directly as the pressure or compression.

3. Each chemical action requires a suitable temperature.

4. In chemical actions which take place over a wide range of temperature the speed of action is much greater at high temperatures. The speed of action is approximately doubled for each rise of 10° C. This gives results surprising to the uninitiated. Say the speed of action is S at 0° C., then:

At 10° C. the speed $= S \times 2 = 2S$; at 20° C. the speed $= S \times 2 \times 2 = 4S$; and at 100° C. the speed $= S \times 2 \times 2 \times 2 \times 2 \times 2 \times 2 \times 2 \times 2 \times 2 \times 2 = 1024S$.

Thus the speed of action may be more than a thousand

times as fast at boiling point as at freezing point. This calculation carried to higher temperatures soon indicates a speed of action almost infinitely great, so that, providing there is any large mass of the ingredients intimately mixed, then the chemical change takes place throughout the whole mass almost at the same instant, thus providing an explosion.

Any chemical action that gives out heat freely will tend to explode. This is what happens in an internal combustion engine cylinder. The spark ignites a portion of the compressed gases before the piston reaches the top dead centre. Streams of explosive flame dart out from the spark-plug points in fan-like rays, causing the temperature of the whole cylinder full of gas to rise to some 200° to 300° C., and as the valves are closed the pressure instantly rises with the temperature. This constitutes the first phase of the explosion and it takes place in the brief space of time while the crank moves from sparking position to top dead centre. The gas is now in the true explosive state; for the increases of temperature and pressure have each independently raised the speed of chemical action, so that the second phase or explosion proper then takes place—the whole of the chemical change taking place almost instantly.

For the sake of completeness definitions of an atom and a molecule are given here; but readers not familiar with them already should also refer to a textbook on chemistry.

An *atom* is the smallest portion of elementary matter that can enter into or be expelled from a compound during a chemical action.

A *molecule* is the smallest portion of a substance that can exist in a free state. Molecules are aggregations of atoms. Thus a molecule of sodium chloride ($NaCl$) is composed of one atom of sodium combined with one atom of chlorine.

When elementary substances such as carbon and oxygen combine to form a third substance, such as the gas carbon monoxide, they always do so in fixed and definite proportions. This is often called the "law of definite proportions."

Thus 12 parts by weight of carbon always combine with 16 parts by weight of oxygen to form 28 parts by weight of carbon monoxide. If, however, the elementary substances can, as in this case, combine to form more than one compound, then, if the amount of one elementary substance is kept constant, the varying amounts of the other (combining with it to form different compounds) are in the relation of integral numbers, usually small. Thus when carbon and oxygen combine to form carbon dioxide, if the amount of carbon is kept at 12 parts as before, then the amount of oxygen combined with it will be 32 parts—the amount of oxygen in the dioxide being just double the amount in the monoxide. This illustrates the law of multiple proportions. There is no possibility of carbon combining with oxygen in proportions of, say, 12 to 17 or 12 to 25.

When a chemical change takes place there is always a redistribution of the energy contained in the substances. Chemical energy may be liberated as heat, or heat may be absorbed and stored in the substances as chemical energy.

Changes, like the union of carbon and oxygen, which liberate heat are called "exothermal" actions. Changes in which heat is absorbed are called "endothermal" actions. The reduction of carbon dioxide to carbon monoxide is such an action. The same chemical change always absorbs or gives out the same amount of heat.

In thermo-chemistry the unit weight of substance used is the molar weight, that is the molecular weight of substance expressed in grammes; and the unit of heat is the gramme-centigrade calorie, which must not be confused with either the pound-calorie (which is rarely used) or the British Thermal Unit (B. Th. U.).

The Chemistry of Carbon

Carbon is one of the elements and has an atomic weight of 12. It is usually tetra-valent, forming compounds like CO_2, but may be bi-valent, forming compounds like CO.

The chemistry of the compounds of carbon is a very extensive and complex subject. It forms one complete division of the science of chemistry and is called organic chemistry, since most of the substances composing, and produced by, living organisms are compounds of carbon. The diamond is the purest natural carbon and the rarest form of the element. Plumbago or graphite is the next purest and is found in limited quantities. Coal contains some free carbon, but most of the carbon is in combination with other elements. Carbon is also in combination in wood, natural gas, mineral oils, and all the metallic carbonates—notably calcium carbonate and magnesium carbonate in limestone and dolomite.

Coke is manufactured in immense quantities by heating coal until all the volatile matter has been driven off. It is a dense type of amorphous carbon and has a variety of industrial uses.

Charcoal is almost pure carbon and is made by heating wood out of contact with air. The distillation of wood yields products partly gaseous and partly liquid. The gases are mainly hydrogen, methane, ethane, ethylene, and carbon monoxide. The fluids are mainly water, wood spirit (methyl alcohol), acetic acid, and tar (see Table IV, p. 31). These are all compounds of carbon, hydrogen, and oxygen. Nitrogen and sulphur may also be present. Table VI (p. 33) shows the gradual rise in the value of carbon in a series of different fuels.

Carbon and Oxygen

Carbon unites with the oxygen of the air to form two different gases, carbon dioxide and carbon monoxide. When air comes in contact with glowing charcoal carbon dioxide is formed. But with an excess of hot charcoal or a limited amount of air the carbon dioxide is changed or reduced to carbon monoxide.

Carbon dioxide (CO_2) is a colourless, odourless gas one-half heavier than air. It is non-poisonous, but because it cannot support life like oxygen it is sometimes regarded as poisonous. It is the gas set free from soda water, lemonade etc. It freezes to a white snow-like mass which is sold for trade purposes, and gives a temperature as low as —80° C.

Carbon monoxide will be considered in more detail.

Carbon Monoxide

Carbon monoxide (CO) is the main explosive component of producer gas. It is a colourless, tasteless, odourless, and extremely poisonous gas—indeed a dangerous combination of properties. It is produced when the carbon dioxide first formed passes on through a layer of glowing charcoal and is reduced thus:

$$CO_2 + C = 2CO$$

Carbon monoxide burns or explodes with air to produce carbon dioxide again:

$$2CO + O_2 = 2CO_2$$

The chain of actions that produce carbon monoxide and explode or burn it are as follows:

1. $C + O_2 = CO_2$ and 14,647 B.Th.U. are liberated for 1 lb. of carbon used in this reaction

2. $CO_2 + C = 2CO$ and 5847 B.Th.U. are absorbed for 1 lb. of carbon used in this reaction.

3. $2CO + O_2 = 2CO_2$ and 10,247 B.Th.U. will be liberated for each 1 lb. of carbon used from the generator.

It must not be forgotten that for each molecule of oxygen that goes into the generator roughly four useless nitrogen molecules go in also, and when air is mixed with the gases in the cylinder of the engine nitrogen again enters in the same proportion. It is this double dilution with nitrogen which reduces the specific output of power from producer gas driven engines.

In the inner fire zone 5 volumes of air (consisting of 1 volume of oxygen plus 4 volumes of nitrogen) will produce 5 volumes of mixed gases if measured at the same temperature and pressure, of which 1 volume is carbon dioxide and 4 volumes are nitrogen.

The second zone action does not affect the nitrogen dilution. In the cylinder the 2 volumes of carbon monoxide require 1 volume of oxygen and therefore 5 volumes of air. There are already 4 volumes of nitrogen mixed with the 2 volumes of carbon monoxide, so that the final result is 8 volumes of nitrogen and 2 volumes of carbon monoxide. If by some means say 50 per cent of the nitrogen could be absorbed either from the air as used, or from the gas coming from the generator, it would raise the engine output to its normal output of h.p. on petrol fuel. It does not seem likely that this can be done.

Water Gas

When steam is passed over glowing charcoal or carbon a mixture of carbon monoxide and hydrogen is produced:

$$C + H_2O = CO + H_2$$

The resultant gases are both combustible and the reaction is therefore of interest in relation to gas producers. Since nothing need be added to the fire but water to get two useful explosive gases, the use and also the limitations of the water-gas system should be studied.

The action is endothermal. That is, it takes heat out of the fire, and if it is continued freely the fire will be quenched. The action requires a high temperature and before the fire is quenched a large amount of useless steam is formed, which is removed from the gas only with difficulty. The fact of the action being endothermal is an advantage if used in moderation, for the exothermal first action in the gas generator has a 60 per cent margin over the endothermal second action. When the engine is under full load the surplus heat developed may be detrimental to the generator and, at any rate, raises the temperature of the gas to an unnecessary degree. The surplus heat could then be usefully employed generating water gas. If, however, the engine is under light load, the surplus heat may be all required to maintain the working temperature of the generator, which is about 900° C.

Water gas can be usefully employed only to the extent that surplus heat is available in the generator. The water spray must therefore be small if it is to be used constantly, or else it must be used intermittently.

To an extent that varies greatly with conditions, water gas is formed automatically from the moisture content of the air supply and from the moisture content of the charcoal. The moisture supplied in these two ways is sometimes all that can be usefully employed. In up-draught generators especially, moisture in the charcoal tends to be driven off as steam before the charcoal falls to the fire zone, and so fails to produce water gas.

Dry charcoal is the ideal. The water for water gas can then be supplied by some suitable mechanism.

Heat Values in the Plant

In order to develop fully the theory of producer gas it is necessary to trace the various transformations of heat in the different sections of the plant. Since this book is intended for practical men rather than for theorists, it may have suited the Australian reader to have these calculations based on "average commercial charcoal" rather than on pure carbon; but carbon has the advantage that the results so obtained are applicable to all carbonaceous fuels, such as coke, anthracite etc. Owing to variations in commercial charcoal, the calorific value, ash, and moisture content may vary almost from bag to bag. If a chemical analysis were made of charcoal in a perfectly dry state, the absorption of moisture at a later date would alter all the values and percentages for the various elements.

It is because of this that Table VI gives the percentage of carbon, hydrogen, oxygen and nitrogen "excluding ash and moisture," and gives the percentage of ash as excluding moisture. This is a device by which the theoretical chemist saves himself a lot of headaches. He declares the water "out of it." This the engineer cannot do. Commercial charcoal will always have moisture in it, for charcoal will neither keep dry after being dried nor keep wet after wetting unless hermetically sealed. Each change in moisture content of the charcoal causes a change in the percentage value of all of the other constituents, and also a change in the calorific value per pound. It is understood that the Council for Scientific and Industrial Research intends to issue a standard or set of standards for charcoal, and this is urgently required. Had such standards been issued in time, it might have been possible to use them in some manner in connexion with these calculations.* But in the absence of

* A specification has been issued in time to include in Appendix II, page 154.

any recognized standard for charcoal it has been necessary to base these calculations on a chemically pure carbon (that it free from moisture, ash or absorbed gases). In this way we can get rid of an almost endless chain of interlinked variables which would otherwise produce confusion. The calorific value of 1 lb. of chemically pure carbon is given as 14,647 B.Th.U. by Brame,* to whom the authors are indebted for this and many other values in the book. A figure rather lower than this is given in other books but we have preferred to keep to Brame's figures, since his reputation is almost unrivalled in this connexion. The use of chemically pure carbon as a basis of reckoning makes the heat values greater per pound than they would be with commercial charcoal, but since commercial charcoal contains 80 per cent or more of pure carbon, the heat values will also be not less than 80 per cent of those calculated at each stage; they may even be rather higher, since some of the absorbed gases have a higher calorific value than pure carbon. The calorific value of commercial charcoal is given in round figures as 13,000 B.Th.U., that is, about 88 per cent of the value for pure carbon, and such a charcoal would produce heat values, at each stage, about 88 per cent of those for pure carbon.

A certain error would ensue because some of the absorbed gases which contribute to the calorific value would be burnt in the inner fire zone and pass unchanged through the outer fire zone, and they might perhaps also be unaffected by the water gas reaction; but on the whole the percentage of carbon is so overwhelming that for practical purposes the error could be neglected.

While on the subject of approximation, it may be noticed that the heat values are calculated out precisely to the last unit; yet for all practical purposes the figures could be

*Brame, J. S. S. and J. G. King: *Fuels: Solid, Liquid and Gaseous.*

rounded off. For instance, 1039 B.Th.U. could be taken as 1000 B.Th.U., 10,247 B.Th.U. could be taken as 10,000 B.Th.U. etc.

We shall now proceed to trace out the transformations of heat which would take place if the fuel were chemically pure carbon and find the heat values for each 1 lb. of pure carbon consumed in the generator.

The values for a good commercial charcoal should be from 80 to 90 per cent of these values.

Heat Cycle for One Pound of Carbon

1. Air enters the generator filled with hot charcoal; ½ lb. of carbon is burned in excess of air at the end of the tuyere or in the inner fire zone:

$$C + O_2 = CO_2$$

and 7323 B.Th.U. are released.

2. The gas passes outwards and combines with another ½ lb. of carbon as follows:

$$C + CO_2 = 2CO$$

Instead of heat being generated by the consumption of the additional ½ lb. of carbon, the endothermic action absorbs 2923 B.Th.U. One pound of carbon has been converted to carbon monoxide with a net heat output of 4400 B.Th.U. at a temperature of about 1000° C. None of this heat is wanted in the engine. The gas entering the engine should be as near air temperature as possible, so the 4400 B.Th.U. must be wasted. The generator efficiency would be only about 70 per cent.

Part of the heat is required to heat the incoming air to the ignition temperature of the fuel (otherwise the fire would go out); part is required to make good losses by radiation from the sides of the generator; and part is necessarily carried off in the gas to the cooler. The water gas reaction is employed to use the waste heat of the gas as far as possible. If it were possible to get an endothermal action to take place at normal air temperature or lower, it would be then possible to use all the surplus heat, but no such action is known. The water gas reaction itself requires a high temperature, and the gas must be at or near this temperature when leaving the generator, and so heat is unavoidably carried away to be wasted in the cooler.

The most efficient exit temperature at which producer gas should leave the generator to enter the cooler has never yet been determined by any competent authority, so far as the author is aware. There is room for a comprehensive set of experiments to determine this. The lower the exit temperature of gas from the generator the less will be the heat loss in the cooler, but at the lower temperature a reverse chemical action takes place between the mixed gases, resulting in carbon dioxide being reproduced:

$$H_2O + CO = CO_2 + H_2$$

This is undesirable. The temperature at which the physical gain by cooling is offset by chemical loss of energy has not been determined. In practice the generator exit temperature rises and falls with the power output. It is suggested that the optimum should be about 600° C.

The Water Gas Reaction

When steam reacts with hot carbon two different reactions occur. At about 500° C. (930° F.) the action is:

(*a*) $$C + 2H_2O = CO_2 + 2H_2.$$

For each 1 lb. of carbon used in this reaction 2840 B.Th.U. must be supplied. At temperatures above about 900° C. (1650° F.) the action is:

(*b*) $$C + H_2O = CO + H_2.$$

For each 1 lb. of carbon used in this reaction 4320 B.Th.U. must be supplied. This is the desirable reaction since it supplies two gases, both explosive, without the admission of any additional inert nitrogen from the air. It is necessary to keep the temperature high enough to favour this reaction.

In stationary producer gas plants about one-third of the carbon is consumed in the water gas reaction, and since each 1 lb. of carbon so consumed requires $1\frac{1}{2}$ lb. of steam, so each $\frac{1}{3}$ lb. requires $\frac{1}{3} \times \frac{3}{2} = \frac{1}{2}$ lb. steam from all sources. This gives the ratio of $\frac{1}{2}$ lb. of water for each 1 lb. of charcoal without allowance for the ash, unburnt smalls, and dust.

This is the best use of water than can be obtained in stationary plants. In mobile producers a smaller proportion of water should be used. The water gas reaction should be shut off at small power output. It is unnecessary to use an amount much in excess of the one water-two carbon ratio at high power output, so that all things considered, including idling and allowing for water vapour in air and charcoal, an overall consumption of 1 lb. of water or steam fed to the generator for each 4 lb. of charcoal would be satisfactory.

If the water can be varied with the power output above about one-third power, then the maximum ratio of 1 to 2 can be maintained, and the water required can be calculated from the fuel requirements as follows: The fuel consumption should be from 0·8 to 1 lb. of charcoal per b.h.p. hr. Taking the lower figure, this is:

$$\frac{8 \times 16}{10} \times \frac{1}{60 \times 60} \text{ oz. fuel per second}$$

There are approximately 480 drops of water per ounce. Therefore the water required per second is:

$$480 \times \tfrac{1}{2} \times \frac{8 \times 16}{10 \times 60 \times 60} = \frac{64}{75} = 1 \text{ drop per second, approx.}$$

This is the "required" water and includes the water vapour in the air and charcoal. It is approximately 1 drop per second for each h.p. that the engine is developing at the time.

In order to simplify the following calculations it is assumed that one-third of the carbon used from the generator is used in the water gas reaction. This is the maximum that can be converted in stationary plants of the best design and having a fairly constant load. With the fluctuating load of a mobile producer unit rather less water would be used even under load, while for idling the water or steam should be shut off from the generator, thus using a lower average proportion of water to charcoal than indicated in the calculations. In order to conserve heat the steam should be generated by the waste heat of the gas at the exit from the generator or by the heat necessarily lost to the tuyere.

TABLE 1

COMPOSITION OF PRODUCER GAS (WITH STEAM)

Sample	Nitrogen (per cent)	Carbon Monoxide (per cent)	Hydrogen (per cent)	Carbon Dioxide (per cent)	Methane (per cent)	Oxygen (per cent)
1	50	28	15	5	2	–
2	60	28	4	8	–	–
3	55	30	10	5	–	–
4	59·6	32·2	6·4	1·6	–	·2

Heat Cycle Including Water Gas

The amount of carbon burned in excess of air is ⅓ lb.,

$$C + O_2 = CO_2$$

and 4882 B.Th.U. are produced.

The carbon dioxide reacts with ⅓ lb. carbon in the outer zone,

$$CO_2 + C = 2CO$$

and 1949 B.Th.U. are absorbed, leaving a net heat output at this stage of 2934 B.Th.U.

If, then, steam is generated by waste heat and the water gas reactions [(*b*) 66 per cent and (*a*) 33 per cent] are adjusted to use a further $\frac{1}{8}$ lb. of carbon, a further 1275 B.Th.U. are absorbed by the reactions plus the heat required to raise the $\frac{1}{2}$ lb. of steam at 212° F. to the temperature of the reaction (1450° F.). This is about 620 B.Th.U. which, added to the 1275, makes 1895 heat units to be taken from the 2934 available.

This would leave 1039 heat units to maintain the generator temperature and all radiation and convection losses therefrom.

Since the 1 lb. of charcoal consumed contained 14,647 B.Th.U. and the losses are 1039 B.Th.U., this gives a gross loss of:

$$\frac{1039 \times 100}{14{,}647} = 7 \text{ per cent}$$

A generator efficiency of 93 per cent would be required. This is the maximum for stationary plants on steady loads. Mobile plants can give a generator efficiency of 80 per cent or better. These are really excellent results considering the very high temperatures employed.

However, with fluctuating load, the efficiency would be lessened and even if the steam generation could be coregulated with fuel consumption to keep to the ratio $\frac{1}{2}$ lb. of total steam to 1 lb. of charcoal, as used in these calculations, it would almost certainly result in overcooling of the generator at light running or idling output levels. The authors are not aware of any device which satisfactorily correlates the steam requirements for water gas production with the variations in power output, atmospheric humidity, and moisture content of the charcoal, all combined.

Heat Values in the Engine

When no water gas reaction is used the heat units available in the engine are as follows:

For each 1 lb. of carbon used from the generator to produce carbon monoxide 4400 B.Th.U. have been liberated and largely dispersed by the losses from the generator and cooler. Some heat remains in the gas, but the less the better in order to get a good weight of gas per stroke; and this, the sensible heat of the gas, is neglected in succeeding calculations.

The heat generated in the engine cylinder for each 1 lb. of carbon taken from the generator in this case is as per reaction 3 (p. 12):

$$2CO + O_2 = 2CO_2 \text{ plus } 10{,}247 \text{ B.Th.U.}$$

Since 1 h.p. = 33,000 ft. lb. per minute, 1 b.h.p. hour = $\frac{33{,}000 \times 60}{778}$ = 2560 B.Th U. for a perfect or 100 per cent efficient engine.

No such engine has been invented, and the efficiency of small engines on producer gas is not much greater than 25 per cent. Therefore the 10,247 units available from 1 lb. of carbon will give $\frac{10{,}247}{2{,}560} \times \frac{25}{100} = 1$ (approx.), that is, approximately 1 b.h.p. hour per lb. of carbon used.

In the water gas reaction only two-thirds of the carbon is used in the air-carbon reaction, and consequently the heat units available at the cylinder from the carbon monoxide from that reaction are only two-thirds the previous figures, i.e. $10{,}247 \times \frac{2}{3}$ B.Th.U. But since each carbon particle taken up in the water gas reaction (*b*), $C + H_2O = CO + H_2$, re-appears as carbon monoxide yielding $10{,}247 \times \frac{1}{3}$ B.Th.U., the heat units available from carbon monoxide per pound of carbon consumed are just as before, 10,247 B.Th.U., leaving the hydrogen to the advantage of the water gas system.

Since carbon monoxide and hydrogen are liberated in equal volumes from the water gas reaction, and since carbon monoxide and hydrogen have very nearly the same calorific value for equal volumes, therefore the hydrogen has very nearly the same heating value as its co-generated carbon monoxide. No great error is introduced by taking the heat available in the cylinder with water gas as $10{,}247 \times \frac{4}{3}$ B.Th.U. And since in the previous case 10,247 units gave one b.h.p. hour, in this case $10{,}247 \times \frac{4}{3}$ units give $1\frac{1}{3}$ b.h.p hours per lb. of carbon used from the generator, or 0·75 lb. of carbon per b.h.p. hour, equal to 0·79 lb. of dry charcoal per b.h.p. hour.

This shows the theoretical fuel economy resulting from water gas to be about 25 per cent.

The non-technical reader must be careful to distinguish the two virtues of water gas:

1. It saves thermal units and therefore saves fuel by gasifying part of the charcoal without burning, and in doing so extracts heat from the gas which would otherwise need to be dissipated in the cooler; also, by medium of the hydrogen generated, the heat extracted from the generator is made available again in the cylinder. This is the fuel economy virtue.

2. Because the water gas reaction produces explosive carbon monoxide and hydrogen without the use of air there is an increase of explosive gas without further addition of useless nitrogen. The engine cylinder fills with a better gas since it is not so occupied with inert nitrogen, and so the power output increases.

It is often stated that the increase of power is due to hydrogen being a better gas than carbon monoxide. Since the cylinder deals with volumes (and, as stated elsewhere, carbon monoxide and hydrogen give out almost an equal number of heat units in combustion, and require almost exactly the same volume of air for that purpose) it seems that the increased power output is due to the smaller proportion of inert nitrogen, and not to any special virtue in hydrogen as an explosive.

Gas producer operators should guard against the use of too much water or steam in the generator, since such a procedure results in the engine cylinder being partly filled with unconverted steam, with a resultant fall in power output.

As the furnace temperature is lowered by increasing amounts of steam supplied, more and more of the steam reacts with carbon according to reaction (*a*), $C + 2H_2O = CO_2 + 2H_2$, which uses just twice as much steam for a given amount of carbon gasified. The fuel economy is not lessened by this reaction even though part of the carbon is converted to useless carbon dioxide, for an extra volume of hydrogen is generated to replace the equal volume of carbon monoxide lost; but with this reaction the cylinder is again crowded with more molecules of an inert gas, i.e. the carbon dioxide, and so the power output from the engine falls.

Because the extra hydrogen of reaction (*a*) replaces the carbon monoxide of reaction (*b*) and each gives the same number of heat units, it has been possible to regard the whole of the $\frac{1}{3}$ lb. of carbon as being used as by reaction (*b*) in calculating the heat available for power, instead of 66 per cent (*b*) and 33 per cent (*a*) used in calculating the heat absorbed in the generator.

Summary

It may be worth while to restate the salient points that

have arisen in the foregoing chapter on the theory of producer gas. This should be especially helpful to readers who have some difficulty in following the chemical and physical theory.

The five essential facts stated at the beginning of this chapter should be read again as part of this recapitulation.

The generator does effective work without the admission of water or steam.

Water or steam may be admitted in any proportion from zero to the maximum of ½ lb. per h.p. hour under ideal conditions.

Water gas production is an endothermal reaction which absorbs heat, so that water gas can only be usefully employed to the extent that surplus heat is available in the generator.

Water gas saves fuel, safeguards the generator, increases the power output—all at no added cost; so that it should be used as freely as is practicable.

In the foregoing calculations water or steam used included all water entering the generator in any way, including the water vapour in the air blast as well as the water content of the charcoal.

The operator is mainly concerned with how much water or steam to add to the uninvited water entering with air and charcoal. This is a really difficult problem depending on a number of simultaneous variables any or all of which may vary from minute to minute. Many advertised plants claim to add water exactly in accordance with requirements, but the authors are sceptical of all such claims.

There is at cruising speed a wide range between no added water and a quenching volume of water. Any water added between these limits is an advantage. This gives even haphazard devices a chance for useful employment. With the engine temporarily stopped, or even idling, every drop of added water may be regarded as quenching water; so none should be admitted.

Many devices such as the wet bottom (Fig. 6) or the outlet boiler (Fig. 8), while fairly effective in plants running at constant load, give too much steam at idling speeds in a mobile producer plant.

The moisture carried in the air varies with the humidity. If air at about 160° F. is saturated with water vapour by passing through water at that temperature it will carry with it all the vapour that can be usefully employed even if the charcoal is dry.

The humidity of the atmosphere is not detrimental since it remains fairly uniform for hours at a stretch; but damp charcoal is very troublesome, especially in up-draught producers, because immediately after refilling, large volumes of steam are given off which fill the cylinders uselessly. In cross-draught producers there is more chance of the adsorbed water reaching the fire zone; but even so the charcoal will get drier as time elapses since refilling, and annoying adjustments must be made to the added water.

If the charcoal is damp enough no water must be added.

If the air humidity were made artificially high that alone could supply the necessary water vapour. A combination of ordinarily dry charcoal (5 to 8 per cent) and common humidity can combine to supply about half the water vapour required.

Until intricate and costly apparatus is devised to deal with all the variables that affect the volume of added water required for a generator, the operator must judge added water by engine response.

TABLE II

WEIGHT AND VOLUME OF AIR REQUIRED FOR COMBUSTION AT 60° F. AND 30 INCHES PRESSURE

Fuel	Weight of Air Required per Lb. of Fuel (Lb.)	Volume of Air Required per Lb. of Fuel (Cubic Feet)	Volume of Air per Cubic Foot of Fuel (Cubic Feet)	Air/Gas Ratio by Volume (Correct Mixture)
Hydrogen	34·80	448·7	2·38	2·38
Carbon (to CO_2)	11·60	149·4	—	—
Carbon (to CO)	5·79	74·7	—	—
Carbon Monoxide	2·47	32·0	2·38	2·38
Methane	17·4	224·3	9·52	9·52
Acetylene	13·37	172·6	11·9	11·9

CHAPTER III

FUELS FOR GAS PRODUCERS

If one were asked to give specifications of an ideal gas producer fuel, the answer would be—a compact fuel with free burning qualities, free from dust, regular in size, giving a rich gas free from tars and injurious products. It should not "coke" in the generator, should be non-hygroscopic (not absorbing water), should not soil the clothes and person of the operator, and lastly should not have an offensive odour either in its generation or from the engine exhaust.

No such ideal fuel yet exists, but probably the fuel which most closely approaches this ideal is the briquetted charcoal developed in France under the name of carbonite. This will be described later, since its possibilities must necessarily be of interest at such a time.

Charcoal

Of the fuels available in Australia there is no doubt that retort charcoal is the most suitable. It gives the cleanest gas, can be produced in almost unlimited quantities, and is free burning (though this quality varies with the type of wood used and the amount of water absorbed by the charcoal). However it fails badly in three essentials: its bulk is considerable; it absorbs water so freely that if not stored well away from moisture it will take up more than its own weight of water; and, being very friable, it breaks up in handling, creating a coating of black dust which penetrates clothing and hair and may soil the "pay load" of the motor vehicle.

An example is cited of a passenger in a car propelled by producer gas who declared that she "left Adelaide a platinum-blonde and arrived at Melbourne a dusky brunette."

Apart from these defects, charcoal is a very high-grade fuel for producers. It is very reactive and forms a gas of fairly constant composition, relatively free from tars, and not difficult to clean. Its ash content is also low, approximately 2 per cent, but owing to its extreme lightness it is carried away in the gas stream, especially in cross-draught generators, and will rapidly clog the filters and scrubber unless the coarse dust is removed by an ash collector, a centrifugal scrubber, or per medium of a water trap.

According to Klar,* the following properties and behaviour are reliable guides as to the satisfactory quality of charcoal: (i) Black and glistening appearance with a distinct bluish tinge; (ii) the production of a metallic note when struck on a hard object; (iii) freedom from taste and smell; (iv) adherence of small pieces to the tongue; and (v) a transverse face can be rubbed without soiling the fingers.

Further, charcoal of good quality should be easily ignited and should burn quietly without the production of flame or smoke. Free-burning qualities are particularly important for gas producer use and, as stated elsewhere, mallee charcoal stands high in this regard, a single coal burning in the open to a small mound of white ash.

Charcoal produced at a temperature of less than 300° C. has a reddish tinge and is known as "foxy" charcoal. In-

* Klar, M., *The Technology of Wood Distillation* (1925).

dividual pieces of red charcoal are known as "brands" and, since they still retain a percentage of tars, they should be removed from the charcoal when grading and re-burnt in the next batch.

Modern science is rapidly bringing all processes to a stage where the skill of the operator is replaced by a combination of machines and instruments. One ancient art has, however, stubbornly resisted attempts to bring it within this category, namely the art of charcoal burning, which even in modern times depends largely on the skill of the operator rather than on automatic control.

This is due to the number of varying factors which come into play during its production. These factors, which to-day are explained by scientific examination, were controlled empirically, that is by a system of trial and error, by the ancient charcoal burners whose experience was handed down from generation to generation.

Charcoal may be divided into three categories, according to its method of manufacture:

1. Pit or beehive kiln charcoal (called in Europe "forest charcoal").

2. That produced in internally fired kilns of firebrick or steel in which, as in type 1, portion of the charge is burned in order to create the necessary heat to raise the remainder of the charge to the temperature at which exothermic action sets in and completes the carbonization.

3. That produced in externally heated closed retorts, usually of steel, in which the wood is carbonized at low temperatures (approximately 400° C.) without the admission of air.

Carbonization of Wood

When wood is heated the first of its constituents driven off is water, accompanied by an evolution of gases composed mainly of oxides of carbon. As soon as the temperature exceeds 280° C. the oxygen compounds diminish and are replaced by hydrocarbons and hydrogen. This evolution of hydrocarbons continues up to a temperature of 500° C., after which they begin to diminish and are replaced by hydrogen.

Up to a temperature of 280° C., heat must be applied to the apparatus; but after this period an exothermic reaction sets in and the carbonization proceeds almost without any further supply of heat. It is at this stage that, if it were possible to judge it, the heat supply should be cut off either by closing the kiln or, in the case of a retort, by removing the fire. The carbonization is then completed by means of the heat given off in the exothermic reaction.

The figures set out in the following table demonstrate very clearly the important part played by temperature in the carbonization of wood. These results also show that a temperature of 400° C. should be sufficient for the carbonization of wood, and in actual practice the maximum temperature seldom exceeds this figure.

TABLE III

Temperature to which wood was exposed (degrees C.)	The Remaining Residues Contained			
	Carbon (per cent)	Hydrogen (per cent)	Oxygen plus Nitrogen including loss (per cent)	Ash (per cent)
150	47·51	6·12	46·29	·080
170	47·77	6·19	45·95	·098
190	50·61	5·11	44·06	·221
210	53·37	4·90	41·53	·200
230	57·14	5·50	37·04	·314
250	65·58	3·81	28·96	·632
300	73·23	4·25	21·96	·569
350	76·60	4·13	18·44	·613
1020	83·29	1·70	13·79	1·224
1500	94·56	·739	3·84	·664

TABLE IV

DESTRUCTIVE DISTILLATION OF WOOD

Solids	Liquids	Gases
Charcoal containing: Carbon (C) and all the ash of the wood. The composition of the ash varies with the variety of the wood and the type of soil on which it is grown but is mainly salts of Ca, Mg, K, Na, Si, Al.	H_2O, Water. CH_3OH, Wood spirit (Methyl Alcohol). CH_3COOH, Acetic Acid. $(CH_3)_2$ CO, Acetone. Tar. Mixed Compounds.	H, Hydrogen. CH_4, Marsh Gas or Methane. C_2H_6, Ethane. C_2H_4, Ethylene. CO, Carbon Monoxide.

The speed of carbonization has considerable influence on the quality of charcoal obtained, charcoal resulting from slow carbonization being richer in hydro-carbon.

Although the carbon content is governed by the temperature at which carbonization is completed, it also depends on whether the charcoal is allowed to cool in the carbonizing apparatus itself or elsewhere.

Juon has shown that if carbonization is stopped at approximately 400° C. (by cutting off the heat supply) and the apparatus is closed up so that neither entry nor escape of gas can take place, further evolution of hydrocarbons may be observed, accompanied by the disappearance of hydrogen and gaseous oxygen compounds. As soon as the apparatus is closed for the purpose of starting the cooling period, the internal pressure increases, and the gaseous atmosphere surrounding the charcoal is found to contain from 80 to 90 per cent of hydrocarbons. These hydrocarbons are, however, energetically adsorbed by the charcoal, resulting in a gradual diminishing of pressure and an increase of from 5 to 6 per cent in the carbon content of the charcoal produced.

It will be seen therefore that the carbonization of wood resolves itself into four different stages.

1. As a result of the application of heat the moisture present in the wood is evaporated. This process takes place at an average temperature of 180° C.

2. Further heating of the wood to a temperature of from 270° to 280° C. is necessary to bring about the period of exothermic reaction or "auto-carbonization" and results in the liberation of gas consisting almost entirely of carbon dioxide and carbon monoxide.

3. The exothermic period proceeds, concentration of carbon in the charcoal taking place and large quantities of hydrocarbons being produced, while the temperature rises to approximately 400° C.

4. Then comes the cooling period, during which the charcoal is cooled, *preferably in the carbonizing apparatus itself* where there is an atmosphere of hydrocarbons, which are adsorbed by the charcoal, resulting in a higher carbon content.

As an indication as to when these changes take place, the experienced charcoal burner relies on the colour of the smoke. A dense white smoke consisting mostly of water vapour (steam) should be evolved in the early stages. A brown tarry smoke denotes too rapid carbonization. A faint bluish smoke in the early stages indicates local overheating, and at a later stage completion of carbonization.

Wood carbonized at high temperatures leaves a residue containing almost pure carbon. With lower temperatures, however, the carbonization is less complete. Although wood itself contains from 45 to 50 per cent of carbon, seldom more than half this yield is obtained even when the wood is carbonized in specially constructed retorts.

Since the useless constituents of the wood are mostly

removed in burning, charcoal will have a greater heating value than the same quantity of wood. The ash content will, however, be higher. The heating power of an average sample of charcoal gives a value between 10,000 and 13,000 B.Th.U. per lb.

TABLE V

ANALYSIS OF FOUR DIFFERENT SAMPLES OF AIR-DRIED WOOD CHARCOAL

Wood Charcoal Constituents	No. 1	No. 2	No. 3	No. 4
Carbon	84	75·5	78	85·0
Hydrogen	2	2·5	3	1·7
Oxygen	6	12·0	7	3·4
Ash	2	1·0	4	3·0
Moisture	6	9·0	8	6·9

TABLE VI

COMPOSITION AND COMPARATIVE CALORIFIC VALUE OF FUELS

Fuel	Percentage, Excluding Ash and Moisture, of				Ash (per cent)	Water (Air-dried) (per cent)	Calorific Value Compared to that of Charcoal (100) (per cent)
	Car-bon	Hydro-gen	Oxy-gen	Nitro-gen			
Wood	45	6	48	1	1·5	18–20	33
Peat	60	6	32	2	5–20	20–30	43
Brown Coal	70	5	24	1	3–30	15	35–75
Soft Coal	82	5	12	1	1–15	4	75–100
Anthracite	94	3	3	–	1·5	2	90–100
Charcoal	95	1·7	3·4	–	4	6	100
Coke	96	·7	2·5	1	4–11	2	95

Burning of Charcoal in Kilns

In France particularly, large numbers of portable charcoal

kilns are in operation. Although differing in construction they are all identical in principle.

The kilns usually consist of a cylinder, the height of which is approximately equal to the diameter, and with capacity from 1 to 20 or more cubic yards.

Provision is made for the controlled admission of air near the base at a number of points round the circumference, while off-takes are provided again round the circumference and also at the top.

The kiln is built up in sections, varying according to size, shape, and method of fixing; but generally speaking the sections can be carried for short distances with reasonable ease by two men. The wood charging the kiln is stacked more or less vertically, the small diameter pieces being at the bottom and the larger ones at the top. Lighting is effected by pouring about 20 lb. of red hot charcoal down a central chimney, reaching to within 2 feet of the base of the stack. This chimney is removed after about 12 to 20 minutes. Some means of regulating the gas or smoke outlet is usually provided, although all regulation can be made at the air intake. The air supply is cut down to a minimum at the start and is gradually increased to a maximum at the end of the firing period. All air ports on both the windward and leeward sides of the plant should be completely closed half-way through the operation.

Most plants are operated at a temperature of about 400° C. The completion of carbonization normally occurs after 15 to 30 hours, depending on the size of the kiln. At the end of this period all air holes are shut, generally sealed with clay, and the plant is allowed to cool. When thoroughly cold, after about 10 or more hours, the kiln is opened and the charcoal removed.

Pit Method of Burning Charcoal

A convenient pit of approximately 12 by 5 feet and 6 feet deep may be used, preferably with sides lined with sheet-iron to exclude the dirt.

Small wood (branches, leaves etc.) is used to kindle a fire at the bottom of the pit and the logs are added to the fire, the larger logs near the bottom and the smaller diameter ones near the top. The wood should be packed as neatly as possible, a long iron bar being used to push the logs into position, and small pieces being packed into the spaces. Too much wood should not be added at one time. The diameter of the logs should not exceed 6 inches for a pit of this size.

About two hours should be taken to fill the pit, the wood being heaped a couple of feet above the ground level and allowed to burn for about one hour until the whole mass is well alight and the pile has sunk to about 2 feet below the ground level. At this stage the fire should be partially quenched with a few buckets of water and then covered in. Sheet-iron or corrugated iron 8-9 feet in length, supported by cross-bars of iron, serves excellently for covering, the ends being covered with earth and accumulations of ash etc., from previous burnings.

The pit is then left to cool off for from one to two days, care being taken that air vents do not occur. If air is allowed to enter the pit the charcoal may be kept alight and burnt away to ash.

When cool the pit may be uncovered, care being taken that the dirt laid on the iron does not fall into the charcoal and cause trouble at a later stage by burning to clinker in the generator. Any unburnt pieces should be picked from the charcoal and put aside for burning in the next batch. After being broken up to a suitable size (not more than 4-inch cubes) the charcoal should be sieved through about a ¼-inch mesh screen to remove the dust and smaller pieces of charcoal. It is then packed in bags and stored away from the rain.

Though it is far from an ideal method, considerable quantities of charcoal for use in gas producers have been burnt in pits with fairly satisfactory results. When mallee stumps are available this method serves quite well, the roots being easier to pack than log wood; and if they are reduced to approximately 9-inch cubes and the dirt is removed, they yield a compact charcoal of free-burning qualities. The clinker produced is usually soft and friable, though it will vary in type with the kind of land from which the roots are taken.

Burning of Charcoal in Drums

In some parts of Australia charcoal is burned in 40-gallon drums. The procedure is to remove the bottom of the drum by cutting away around the rim. Three or four slots with an axe or cold chisel are made in the top and the filler plugs removed. The drum is inverted and a fire started at the bottom. It is then filled either with wood of approximately the length of the sides of the drum, or, preferably, with sawn billets about the size of firewood "cut blocks."

Air is drawn to the fire through the plug holes and gashes, the drum being propped up slightly to allow the air access. When the wood is well burnt down the bottom is replaced as a lid, and either wired down or fixed with a couple of iron skewers. The drum is then turned over again (the most difficult part of the process).

The air is excluded completely by covering the slits with a damp bag and heaping soil on it. Soil and ashes should also be heaped around the drum at ground level. By this method charcoal can in an emergency be produced overnight; but on account of the small amount produced (about one bag per drum) it can hardly be considered an economical method. Its worst defect is that the charcoal so produced invariably contains pieces that are only partially burnt, the gas from these pieces carrying an undue proportion of tars and pyroligneous acid which cause engine corrosion and gumming of valves. This process cannot, therefore, be recommended.

Woods for Producing Charcoal

Although all woods can be burnt to charcoal, certain kinds are particularly good for use in gas producers. Of these undoubtedly the hardwoods are the best; and, in the same locality, wood cut from the highlands is better than the same type of wood cut on the lowlands or swampy country.

Hardwood charcoal is better because it is more compact and does not produce much dust when travelling. Woods grown on swampy lands, particularly if saline, produce a charcoal yielding a high percentage of clinker.

Briquetted Charcoal

As pointed out elsewhere, briquetted fuels most nearly satisfy the ideal.

At present the demand in Australia for ordinary charcoal is not large, and that for briquetted charcoal is less still. If this demand becomes sufficiently great there will no doubt be special works equipped here for its manufacture.

One of the first questions requiring investigation will be whether a binder is needed, and, if so, what is the best. One type of French briquette is bound by a vegetable compound; another by what is claimed to be flour; another has no added binding at all, relying solely on the tar present in the charcoal.

With briquetting there is little wastage of wood, since practically the whole of the chips etc. are used.

Carbonite

Carbonite is one of the most readily obtainable briquetted charcoals in France. In the manufacture of carbonite the wood is chopped into small pieces and distilled, with the recovery of the tars, acids etc. These are, after treatment, used as a binder for the charcoal. In this way great economy is effected. Approximately 700 lb. of carbonite are made from 1 ton of pine.

The use of the volatiles for binding means that they are present in the briquette to an extent of 7-8 per cent. This adds considerably to the calorific value of the fuel.

Carbonite is made in ovoid briquettes of approximately ½ oz. weight. These briquettes have a specific gravity of greater than one, which is four times that of charcoal. This would allow the use of a generator a quarter the size of that necessary for charcoal, or would give four times the range for the same capacity.

Carbonite briquettes are much less friable than charcoal and produce very much less dust. The compactness of the briquettes does not appear to be detrimental to their reactivity, but rather introduces a desirable feature. If left to burn on their own, carbonite briquettes become covered with a coating of ash which is easily blown away, leaving a white-hot gas-producing surface. This allows of very quick pick-up after long periods of idling or light running.

Perhaps the worst feature of briquetted fuels is the burning of the binder, which usually produces large amounts of ash. If the volatiles are used as a binder the resulting gas will contain a large proportion of tars and acids which have a harmful effect on the engine.

Shortly after the last war there was constructed in Germany a plant for the briquetting of charcoal obtained from sawdust—a waste product of the saw mills. This works was, however, forced to close down owing to technical difficulties, the worst of which was the friability of the briquette produced. Difficulties were also encountered in the burning of the sawdust; this had to be burnt on chain grates which were a continual source of trouble.

TABLE VII

DATA CONCERNING GASEOUS FUELS AT 60° F. AND 30 INCHES PRESSURE

Gas	Formula	Weight per Cubic Foot (Gramme)	Volume per Lb. (Cubic Feet)	Calorific Value per Cubic Foot (B. Th. U.)
Oxygen	O_2	0·0844	11·85	—
Nitrogen	N_2	0·0742	13·50	—
Air	—	0·0764	13·10	—
Carbon Dioxide ..	CO_2	0·1168	8·56	—
Carbon Monoxide	CO	0·0739	13·50	323·5
Hydrogen	H_2	0·0053	188·70	325·2
Methane	CH_4	0·0422	23·7	1010·00
Acetylene	C_2H_2	0·0703	14·23	1475·00

Calibration of Charcoal

It is important that charcoal for gas producer use should not contain big lumps which allow vaults to form in the fire. This is particularly important in a cross-draught generator. One maker recommends that the charcoal should never be bigger than the diameter of the tuyere hole of the generator for which it is prepared.

OTHER FUELS

Although charcoal is the principal fuel used in gas producers in this country, a number of other fuels will be briefly dealt with in order to create a better understanding of the subject.

Coal

Coal has been formed by the bacterial decomposition of the plant remains of past ages; these remains, being buried in the earth in an absence of air and under high compression, have undergone slow mineralization. Various stages in the conversion of vegetable matter into coal are marked by the substances known as peat, lignite, bituminous coal, cannel coal, and anthracite.

Lignite

Lignite or "brown coal," as it is often called, differs from peat and coal in that it is mainly composed of "fossil wood." There seems little doubt that this would in the course of time have been converted into coal. It has a a heating value of from 9000-13,800 B.Th.U., but the amount of moisture and ash generally present makes it an inferior fuel. It is, however, of particular interest to us in Australia, since it is the only coal available in briquetted form.

The State Electricity Commission works at Yallourn, Victoria, produce large quantities of briquetted brown coal, both for household and industrial uses. Some experiments have been conducted by the Commission, in conjunction with at least one Australian manufacturer, into the use of brown coal briquettes in gas producers. But the high proportion of hydrocarbons present produces large quantities of tar and necessitates the use of special furnaces of the dual-combustion-chamber type, the gas passing through a bed of incandescent charcoal or coke in which the hydrocarbons are "cracked" and the tarry compounds reduced to a combustible gas.

It is understood that the Commission is experimenting with a briquette char- which may in the future overcome these disabilities and ensure a constant supply of a highly suitable fuel.

Bituminous Coal

Bituminous coal is the commonest type of coal. It contains a large proportion of volatile constituents and therefore gives a large proportion of gas when heated. It burns with a bright and long flame. It has a heating value from 14,500-15,300 B.Th.U. On account of its high tar content it is not suitable for use in a gas producer unless first carbonized to coke.

Coke

Coke is formed by the destructive distillation of coal, in much the same manner as charcoal is produced from wood. It consists chiefly of carbon (about 80 per cent) and still contains some of the original nitrogen of the coal, together with about 2 per cent of sulphur.

Anthracite

Anthracite is the least widely distributed of the coals and one of the most valuable. It is a very hard substance, containing a large proportion of carbon and but little volatile matter. Consequently it gives a very intense local heat when ignited fully and produces little smoke or ash. It has been extensively used in England for gas producers.

Wood

The abundance of wood, its comparatively rapid growth and reproduction and the ease of obtaining supplies make it a very cheap fuel. The fundamental chemical constituents of wood are cellulose ($C_6 H_{10} O_5$) containing 44 per cent of carbon, and the so-called incrusting substance which is richer in carbon and is known as "lignin." The exact composition of lignin is not known but the substance is possibly made up of several carbon compounds.

Under the most favourable conditions wood will be of a low value as a fuel because it contains only some 80 per cent of combustible substances, which have a low heating value. In addition the large amount of moisture present demands much of the available heat for its vaporization. On the other hand, it is easy to ignite and can be burnt without difficulty.

When burnt in a gas producer wood produces an excellent gas, but certain complications necessary in the generator and in the cleaning apparatus, together with its large bulk, justify the trouble and cost of burning it to charcoal, rather than using it as it is. A description of a wood-burning plant is included in Chapter XII.

CHAPTER IV

THE GENERATOR

THE function of the generator is the production of a combustible gas for use in the engine. Its efficient working, therefore, is of primary importance to the producer as a whole.

The gas is produced by burning a suitable fuel such as charcoal, anthracite coal, coke, or, under certain conditions, wood, in the generator with a limited supply of air.

UP-DRAUGHT GENERATOR

The simplest form of generator is shown in Fig. 1 and consists essentially of a cylindrical or rectangular container to which has been fitted a removable air-tight lid to permit the introduction of fuel which is burned on a grate near the bottom of the generator. A door for the removal of ash and one or more holes for the entry of air are provided below the grate. The gas outlet is placed near the top of the container and connects through the scrubbers to the engine manifold.

The suction of the engine causes air to be drawn up through the grate. Since a fire always burns at the point where air is admitted to the fuel, gas will be formed above the grate (as described in Chapter II) and will pass upwards through the fuel to the gas outlet.

This type of generator is known as an up-draught generator. Though of simple construction and requiring some modification in practice, it is essentially the basis of all generator design. When fitted with a refractory lining for the furnace (the section where the fire actually burns), and a revolving or movable grate, it has been largely used for stationary plants and, with certain other modifications, for motor vehicles.

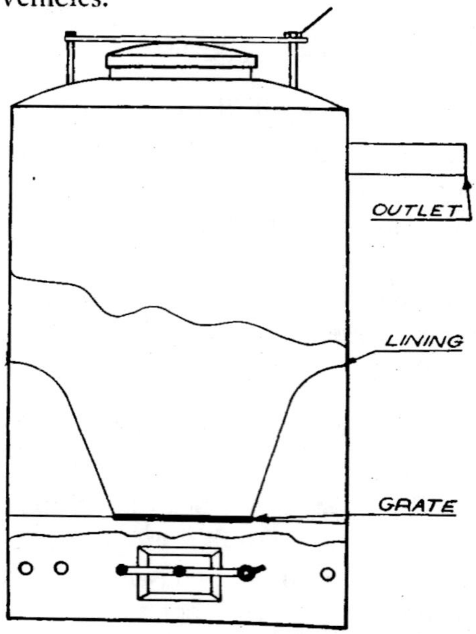

Fig. 1. A SIMPLE TYPE OF GENERATOR

The gas produced by an up-draught generator contains a proportion of tar which it collects as it passes up through the heated but unburnt fuel.

Down-draught Generator

A down-draught generator is one in which the air is admitted at the top of the generator, preferably through a centre pipe carrying a water-cooled or refractory nozzle, passes down through the grate at the bottom of the generator, and from there is drawn to the cooler and scrubber.

The advantage claimed for this type is that the combustible gases are taken off at the hottest part of the fire and so contain less of the volatiles, such as tar and wood-naptha. Fig. 2 shows a sketch of a simple generator operated on the down-draught principle.

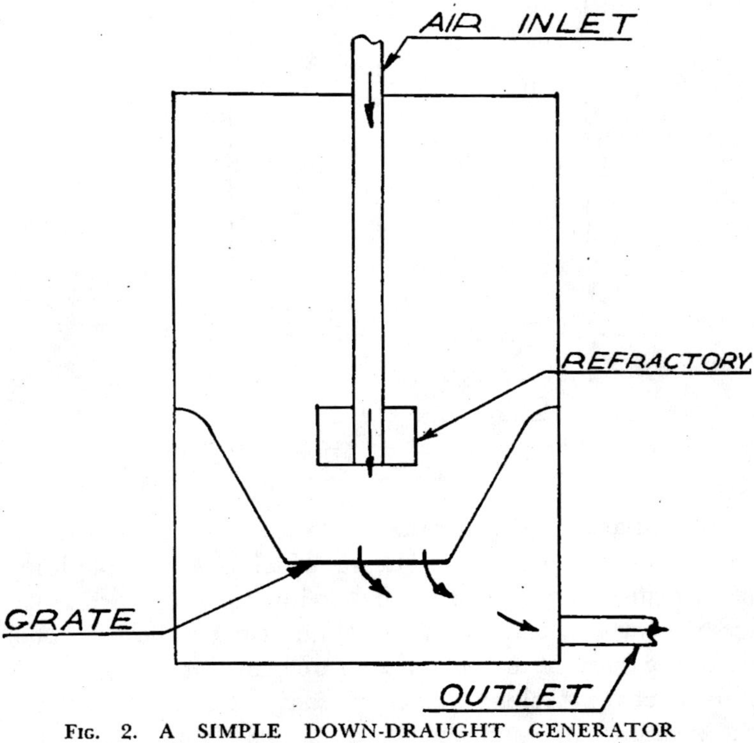

Fig. 2. A SIMPLE DOWN-DRAUGHT GENERATOR

Cross-draught Generator

Of late years a much more suitable generator for motor vehicles has been produced which, instead of using a grate, employs what is known as a tuyere or nozzle (see Fig. 4). In this case the path of the gas through the furnace is in a horizontal direction, instead of vertically as in the previous types. This generator, of which the French plant of Gohin-Poulenc was one of the earliest examples, is known as a cross-draught type (see Fig. 3).

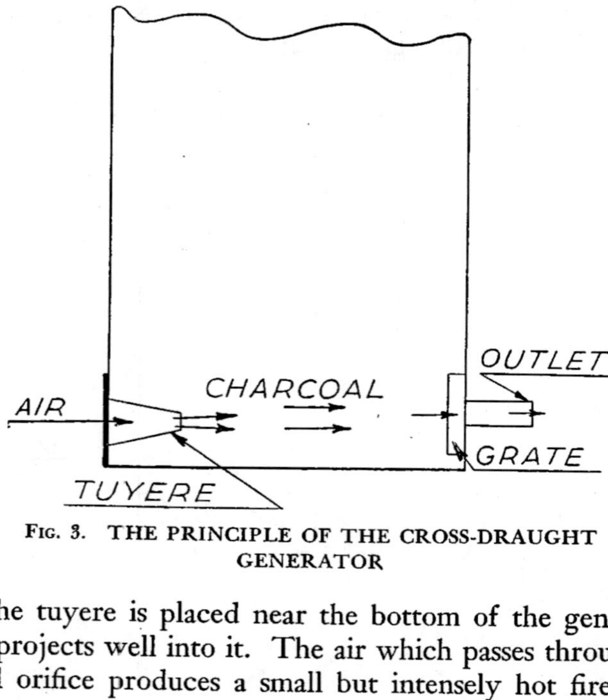

FIG. 3. THE PRINCIPLE OF THE CROSS-DRAUGHT GENERATOR

The tuyere is placed near the bottom of the generator and projects well into it. The air which passes through its small orifice produces a small but intensely hot fire zone in the centre of the generator. Thus around the fire zone there is a body of lower temperature charcoal which serves to protect the steel walls from excessive heat and eliminates the necessity of employing a refractory lining. The gas outlet is placed directly opposite the tuyere so that the gas produced is freer from tar, owing to the fact that it passes through a smaller amount of partially heated fuel than in the case of an up-draught generator. The gas outlet is covered with a grate or screen to prevent pieces of the fuel being carried away in the gas stream.

To prevent the tuyere being destroyed by heat, provision is made for it to be cooled by circulating water through it. This may be done by a connexion to the

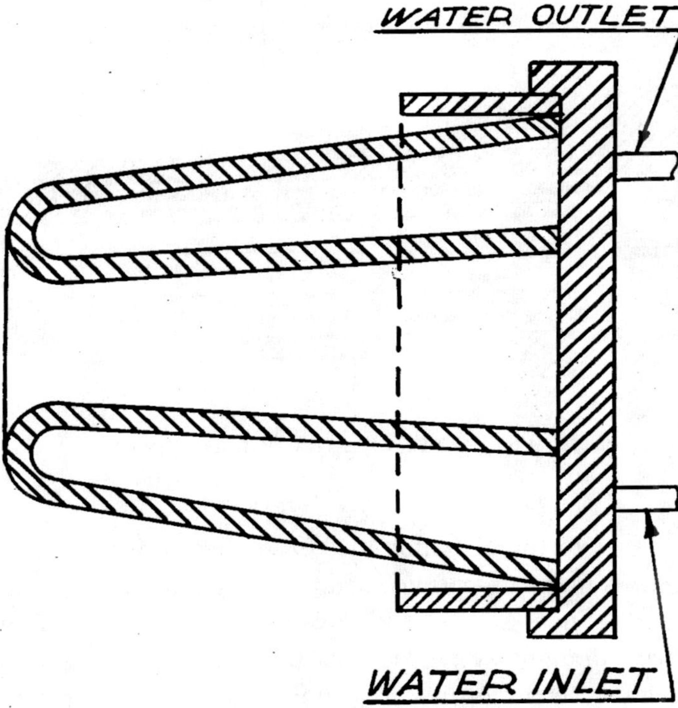

Fig. 4. A WATER-COOLED TUYERE

radiator of the vehicle or by means of a special water tank provided for this purpose.

Fig. 5 shows a novel type of air-cooled tuyere which has been developed in France by Messrs Sabatier. As can be seen, this consists of two concentric tubes arranged in such a manner that the incoming air itself is the cooling agent. Thus the amount of cooling varies with the velocity of the air stream which in turn controls the temperature of the fire. In this system all danger of destruction due to a blockage or break in the water supply is avoided. However, the resistance to the passage of the air through it is considerably more than that of an ordinary tuyere, since it has to traverse a far more difficult passage.

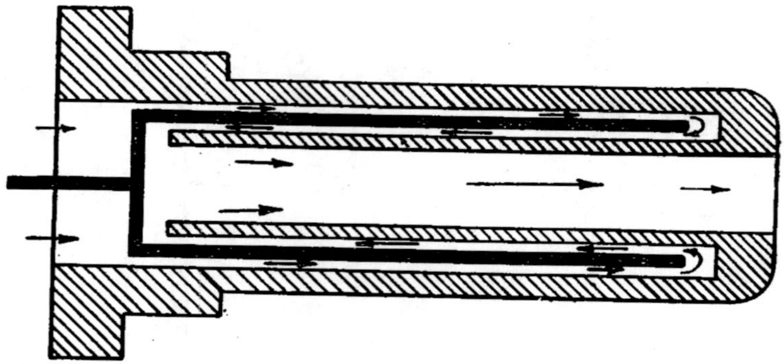

FIG. 5. THE SABATIER AIR-COOLED TUYERE

OTHER TYPES OF GENERATORS

In addition to the orthodox up-, cross-, and down-draught generators there are certain modifications and combinations of these. Such is the Brush-Koela plant which employs the cross-draught system for starting and the up-draught for regular operation. It is claimed of this arrangement, which is named the duo-draught, that vastly quicker starting can be obtained together with more economical operation.

Among the more novel generators which have been designed is one developed in the mines of Anzin. This, a stationary plant, has a refractory lining into which two tuyeres are built and inclined downwards at 45° towards the centre of the generator, the gas outlet pipe being at the bottom.

Admission of Water

As seen in Chapter II, water is used for two purposes—to control the temperature of the fire and to generate a richer gas. The means of admitting this water varies with the different makes of plants, but in nearly all cases it is let into the air stream in the form of steam, the water being brought to this state by means of the heat of the hot gases or of the fire itself.

The simplest means of steam admission in an up-draught generator is by allowing the cinders to fall through the grate into water contained in the bottom of the ash box, so creating steam which mixes with the incoming air and passes to the fire (see Fig. 6).

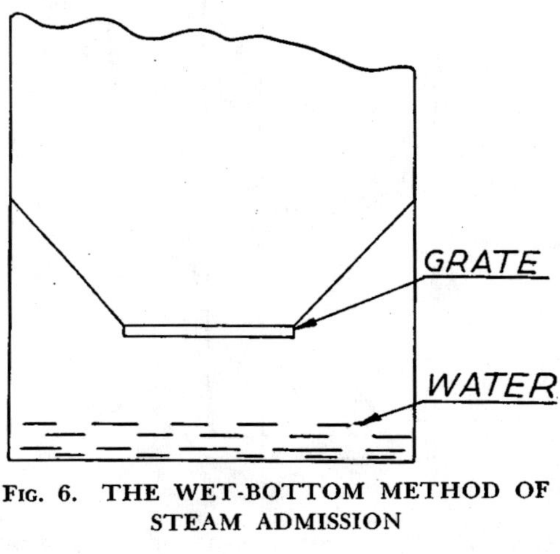

Fig. 6. THE WET-BOTTOM METHOD OF STEAM ADMISSION

This method, though simple, is not subject to control by the operator. Moreover, bumpy road surfaces cause a considerable amount of cinders to fall into the water, perhaps at the most unsuitable times so far as the furnace temperatures are concerned.

Other methods of application which apply to various

E

types of generators use a manually or automatically controlled water drip feed, admitting water to the heated portions of the furnace walls, grate, or tuyere as the case may be, and so creating steam which is carried into the furnace by the air stream.

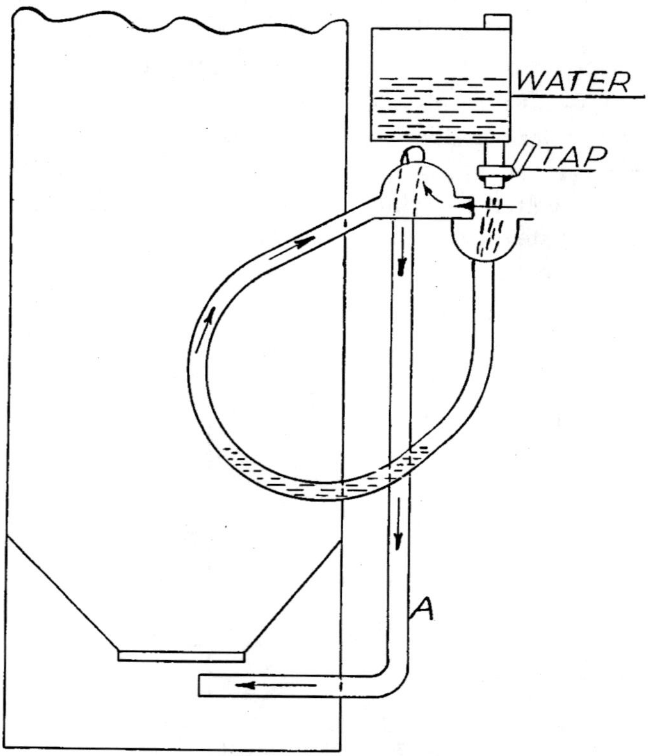

Fig. 7. A DRIP-FEED ARRANGEMENT FOR ADMITTING STEAM

A number of means of automatic control have been adopted, such as diaphragmatic control by means of furnace pressures, harnessing the steam control to the accelerator pedal, etc.; but all of these involve certain complications which prevent their general adoption.

Figs 7 and 8 show two methods of admitting water. The latter cannot be so easily controlled by the operator. The apparatus shown in Fig. 7 is made up of a tank from which water drips into the mouth of a U-tube, one side of which is within the generator. In the lower sections of this tube there will be a pool of water which is being heated by the fire. Steam is given off and passes up the tube into a beehive cone, where it mixes with the incoming air and passes along the tube A, which is let into the generator below the grate. In this manner a mixture of air and steam is supplied to the fire.

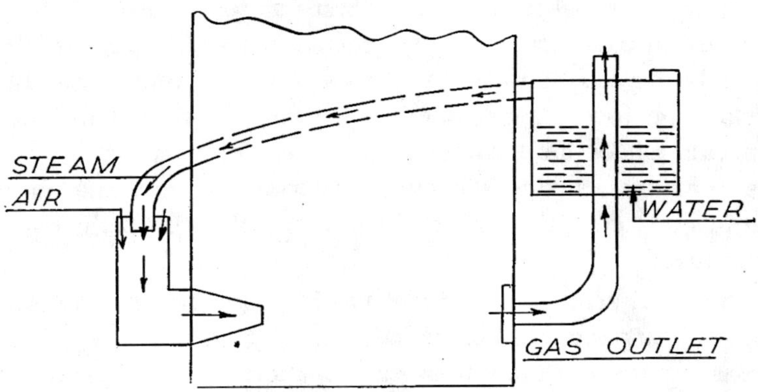

FIG. 8. MEANS OF ADMITTING STEAM TO CROSS-DRAUGHT GENERATOR

The system shown in Fig. 8 is an attempt at an automatic means of steam admission, the steam being produced by means of the hot gases from the generator. The advantage of this arrangement is that no steam will be admitted until the fire is hot enough to generate gases that can boil water and so give off steam, which is admitted with air to the tuyere. A more automatic arrangement could be arrived at by fitting a ball and cock valve, which would make allowance for the water boiled away.

In France a number of the principal manufacturers do not use steam, but in Australia most makers do. Possibly the reason is that, owing to the wetter climate, charcoal in France generally has a higher water content than in Australia, and the advantages of steam admission are correspondingly decreased.

Air Cooling and Preheating

As previously seen, there is no necessity to use a lining in the case of a cross-draught generator, but this does not always apply to the up- and down-draught types.

The older makes of these generators used mainly a lining of fire-brick or some similar refractory substance, which was both costly and bulky. Nowadays this lining usually takes the form of an inner shell which is cooled by the circulation of air between it and the outer casing. This is done by utilizing the engine suction to draw the cool atmospheric air through this space and thence up through the fire.

Such an arrangement has a twofold use, for in addition it preheats the air before admitting it to the fire, thus assisting the production of carbon monoxide. A diagram of a generator using this system of cooling is shown in Fig. 9.

With the correct conditions prevailing there is a tendency for minor explosions to occur in all generators. Flame is blown through all inlets and outlets, including the air inlet. There is a grave danger of fire being caused in this manner unless some precaution is taken.

For this reason the end of the tuyere is usually covered and the air inlet pipe let into it. This safety tube, as shown in Fig. 10, extends up along the side of the generator and is provided with a gauze cap.

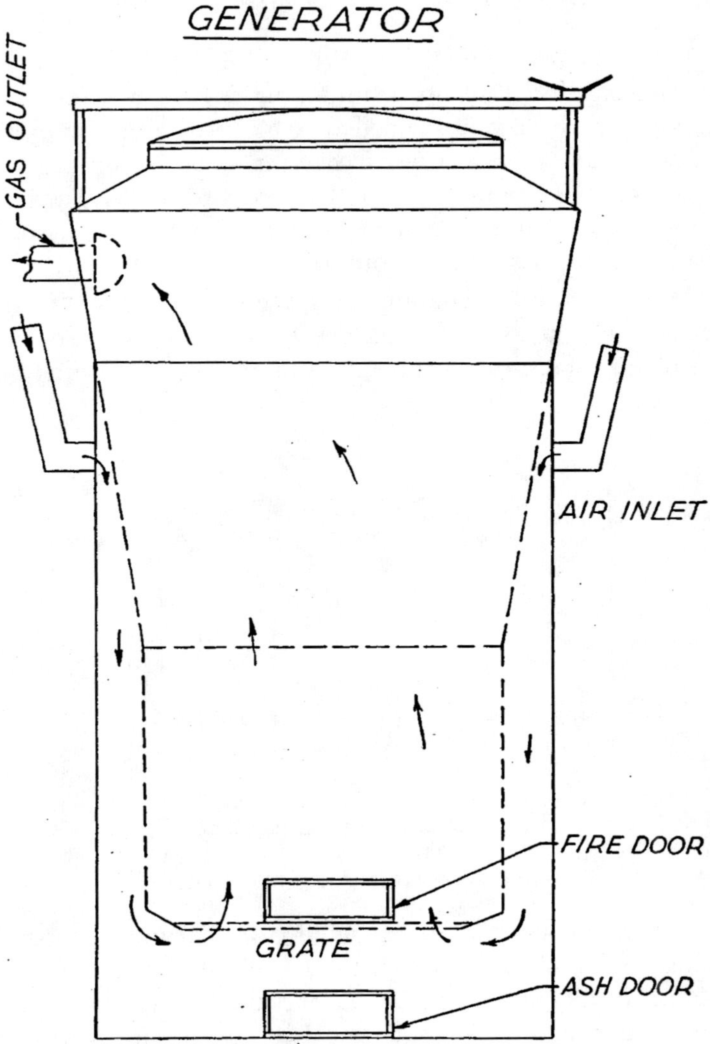

Fig. 9. AN UP-DRAUGHT GENERATOR, SHOWING THE METHOD OF PREHEATING THE AIR AND COOLING THE FURNACE WALLS

The Shape of the Generator

A review of the plants manufactured here and overseas shows that there is a decided tendency towards the cylindrical shape. The reason is that this type is cheaper and easier to manufacture and is not so readily damaged by explosions. On the other hand it occupies more space than would a rectangular one of the same capacity.

With a view to economy of space a number of "freak" shaped generators have been designed; but these introduce many manufacturing problems, and are therefore likely to be costly.

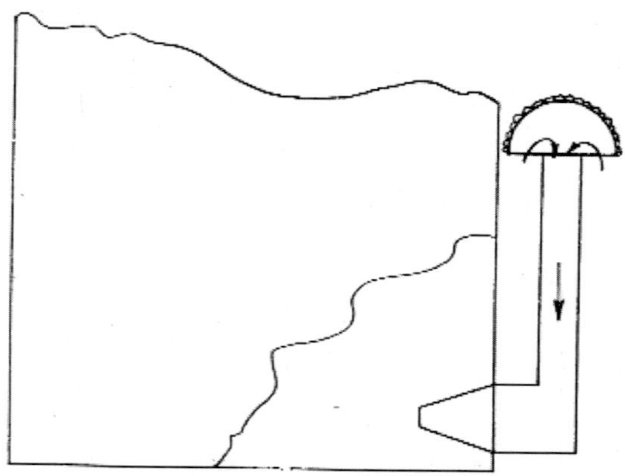

Fig. 10. AIR INLET TUBE ON A CROSS-DRAUGHT GENERATOR

Besides generators constructed for burning charcoal, special ones have been built which burn wood and in which arrangements are made for the separation of the tars. Although few have yet been made in Australia, a description of wood-burning plants will be included in Chapter XII.

CHAPTER V

THE SCRUBBER

As its name implies, the scrubber cleans the gas before it enters the engine.

There are several reasons why this cleaning is necessary. As the gas leaves the generator it carries with it a certain proportion of dust, ashes, water vapour and tarry substances which have been derived from the fuel. If these were allowed to enter the engine the dust and ashes would cause excessive wear; the tar, excessive carbon deposits, and in exceptional cases the seizing of the engine; and the water vapour, rusting of the valve seats and sludging of the oil.

Before the introduction of the suction gas producer, scrubbers or cleaners were in use in the manufacture of town gas. These were later adapted for use with producer gas.

These exceedingly simple scrubbers were usually of the "coke tower" type. Tall, vertical containers are packed with coke, through which water percolates in opposition to the flow of gas, which is admitted at the bottom of the tower and drawn off at the top. If made sufficiently large, this type forms a very effective scrubber as well as a cooler for the gas. But the considerable size necessary as well as the difficulty of providing a supply of water renders it unsuitable for motor vehicles. It is still used, however, in stationary plants.

Many stationary plants in which the ordinary coke

scrubber is insufficient are provided with a tar extractor, such as the National in which, after the gas impinges on plates carrying a series of flanges, it passes through a final sawdust scrubber.

Some manufacturers employ liquids such as water and oil to clean the gas, while others use only dry substances, such as sisal hemp, coke, fibre, felt etc. The latter are referred to as dry scrubbers, the former as wet scrubbers.

Wet Scrubbers

In wet scrubbers the gas is drawn by the engine suction through pipes or beneath baffles projecting below the surface of the water. Serrated edges and in some cases coarse gauze are used to split the gas stream and prevent large bubbles from carrying foreign matter through the liquid. In some plants oil is also used on account of its great affinity with tar.

Water as a Cleaner

Water is not a highly efficient cleaner, since it possesses no special powers for the elimination of tar. But although it is not generally favoured, water provides a cheap and effective means of removing dust and ash as well as of cooling the gas. Further, and this is of particular importance, water provides a simple and effective protection against explosions in the scrubber unit caused by burning gas from the generator.

Oil as a Cleaner

Oil possesses good properties for cleaning gas if passed through it at moderate velocity, when it acts as an efficient tar, dust and ash remover. This will necessitate the use of a large cross-sectional area container.

There are numerous ways of applying the oil as a cleaner. The gas may be bubbled up through an oil container or sucked through a filtering medium, such as felt or gauze which is soaked with oil. Still another method consists of bubbling it up through water, on the top of which is a film of oil.

Chevalet has patented the use of coal tar oils for the removal of tar and has shown that heavy paraffin oils will only collect the particles mechanically, while anthracene and creosote oils actually dissolve the tar.

In the operation of an oil scrubber a very interesting incident occurs which emphasizes its good cleaning properties. It will be found that after several days' use the oil in the scrubber will have greatly increased its volume owing to the addition of the impurities removed from the gas.

Dry Scrubbers

In view of the difficulty of carrying liquids on a motor vehicle plant, manufacturers developed the use of dry scrubbers in which coke, sisal hemp, wood-wool, chopped cloth, sawdust, felt, horsehair and other similar substances are packed in containers or in the pipes of the cooling system.

These substances are very efficient cleaners up to a certain point; but there is a danger of their becoming contaminated with the tar and dust, which, in addition to offering considerable resistance to the passage of the gas, may be carried by it into the engine. In other words, like a dirty duster, it soils more than it cleans.

Felt as a Cleaner

Felt has excellent cleaning properties, but tends to become clogged with dust, producing a rapid rise in the back pressure. For this reason it is generally used as a final cleaner to remove impurities that have passed through the primary scrubbers. A large area is desirable.

Sisal Hemp as a Cleaner

In Australia one of the most extensively used cleaners is sisal hemp. To be effective the sisal should be wet and preferably used in combination with some other cleaner, such as felt. As large a cross-sectional area as possible should be used without making the scrubber cumbersome.

Horsehair as a Cleaner

Horsehair provides an excellent substitute for sisal, having superior heat-resisting qualities and less tendency to "mat." When soaked with oil and compacted in layers it makes an effective cleaner. It is, however, very much more expensive than sisal, except on a farm.

Sawdust as a Cleaner

Although not much used in Australia, sawdust has figured prominently in a number of oversea scrubbers, particularly in those for stationary plants. It has reasonable cleaning properties and is cheap, but offers a fairly high resistance to the passage of the gas. A shallow bed of large area would appear to be a good cleaner.

Steel Wool as a Cleaner

Another cleaner which has received a limited amount of use is steel wool. With this there does not exist the same danger of clogging, but it has been found that after several days' use a considerable quantity of the wool has been missing—presumably having been carried by the gas into the engine, where the damage it would do can readily be imagined.

All the other substances mentioned, as dry scrubbers, possess cleaning properties similar to those just described, and can be used with a reasonable amount of success.

Filter Scrubbers

Just as solid impurities may be removed from water by filtering, so impurities may be removed from a gas by passing it through a filtering cloth of fine mesh. Often an ordinary dry scrubber is referred to as a filter, but although the action is much the same, for identification purposes only those scrubbers in which the gas passes through a cloth should be known as filters, the others being referred to merely as dry scrubbers.

Filter scrubbers, though varied in design, all employ some kind of metal frame over which the cloth is tightly stretched. Almost any cloth could be used in a filter; but for good results the cloth should have certain properties —it should not stretch or shrink during use, should not be of too fine or too coarse a mesh, and should not be too easily affected by heat.

Removal of Dust from the Filter

A filter cloth should not be fitted horizontally, since in such an arrangement there is an upper surface from which the dust cannot fall. If the cloth is placed vertically, it is impossible for the dust to accumulate on the surface and so cause a blockage, because its own weight and the vibration caused by travelling will make it fall to some lower surface from which it can be removed.

Burning of Filter Cloths

Perhaps the worst feature of a filter scrubber is the possibility of its being destroyed either by too hot a gas or by an actual explosion occurring in the scrubber. For these reasons a filter scrubber must always be placed in the line after the cooler.

Removal and Destruction of Tar

There are two methods available for dealing with the

question of tar: (i) its removal from the gas, and (ii) its destruction in the generator by means of a "reducing zone" (as shown in the wood-burning plant in Fig. 29).

Coke Scrubbers in Stationary Plants

Since water forms the most convenient medium for cooling the gas, cooling and tar removal are effected simultaneously by the use of coke scrubbers through which water is passing.

Since water and tar do not mix, the water in no way helps in the removal of tar, merely acting as a cooling agent. In scrubbers for motor vehicles, in which the cooling is performed by air, the water is eliminated, leaving a scrubber containing only coke, the cleansing action of which is due partly to the constant change in the direction of flow of the gas, caused by the irregular coke surface.

The tar already deposited in the coke, however, plays the most important part in its further removal from the gas, for surfaces flooded with tar are very effective scrubbers when the gas is forced into intimate contact with them.

Removal of Tar by Condensation

If steam is rapidly cooled against a cold surface, water is deposited on this surface. Similarly, if producer gas is rapidly cooled it deposits its tar as a thin dry fog and, if more slowly cooled, as heavy wet globules which are more easily removed.

It is difficult to obtain sufficient cooling to produce this condensation if air is the agent used, because the large volume of gases passed through the cooler soon heats it to a temperature above that at which tar condenses. However, a certain amount of condensation is effected in all cooling apparatus and provision should always be made for the removal of the condensate.

Gassification or Destruction of the Tar in the Fire

Instead of removing the tar from the gas, there is much to be said in favour of converting it in the furnace into a permanent gas, thereby adding its calorific value to that of the gas and eliminating the necessity for much of the scrubbing.

In certain French plants using wood or bituminous coal, provision is made for leading the gas (which contains a higher percentage of tar than charcoal gas) through a zone of incandescent charcoal or coke, where the tar is converted into hydrocarbons which pass with the other gases to the engine.

Tar Extractors for Stationary Plants

A number of stationary plants employ static tar extractors, among these being the Burstall and the Pelouoz and Audouins, as described in Brame's and King's work on fuels.*

The principle upon which the Pelouoz and Audouins tar extractor works is that a sudden change in the direction of flow of small gas streams issuing from round holes in one series of plates and impinging on plates correspondingly slotted, on which there is tar from the gas previously passed through, causes fine globules to be collected from the fresh gas.

In Professor Burstall's static washer a large number of wires, fixed only at their upper ends, are suspended in a rectangular tank. Water is forced against the top of these wires by holes in transverse pipes. There are 117 sections of wires with 58 and 59 wires alternately in each section. This arrangement ensures that when the gas is passing through the washer it comes into intimate contact with the

---

* **Brame, J. S. S. and J. G. King, *Fuels: Solid, Liquid and Gaseous.***

wet wires, because a wire in one section is in line with a gap between two wires in the sections before and behind.

Centrifugal Scrubbers

All the cleaners so far described are static. We shall now go on to consider another type—namely the centrifugal scrubber. Centrifugal cleaners can be divided into two classes—those in which the baffles or vanes remain stationary, and those in which the vanes are revolving. Because the latter require power to drive them they are not generally favoured for motor vehicles; but practically all gas producer plants have one of the former incorporated in them under the name of a dust extractor or cyclone.

The cyclone consists essentially of a container in which a number of vanes are arranged in such manner as to cause the gas rapidly to change its direction of travel. This rapid change of motion causes the heavier dust particles to be thrown out from the gas and fall to the bottom of the container, from which they are periodically removed by means of a door provided for the purpose.

Of the centrifugal cleaners of the fan type the Crossley and Burstall cleaners are the most generally used.

The Crossley Centrifugal Cleaner

In the Crossley centrifugal cleaner (see Fig. 11) the gases are directed to the centre of the fan on the inlet side and can pass to the exit only by travelling outwards on the inlet side of the casing, and back to the centre exit to the outlet pipe. Water is made to pass through the exit and, together with the separated tar, is thrown against the casing, to drain finally to the tar sump.

The Burstall Cleaner

Burstall's centrifugal cleaner consists of a 24-inch rotor constructed of wires. This is enclosed in a casing arranged in such a manner that the gas enters it near the centre and leaves at the outside, thus coming into intimate contact with the tar-covered wires which are revolving at approximately 1800 r.p.m.

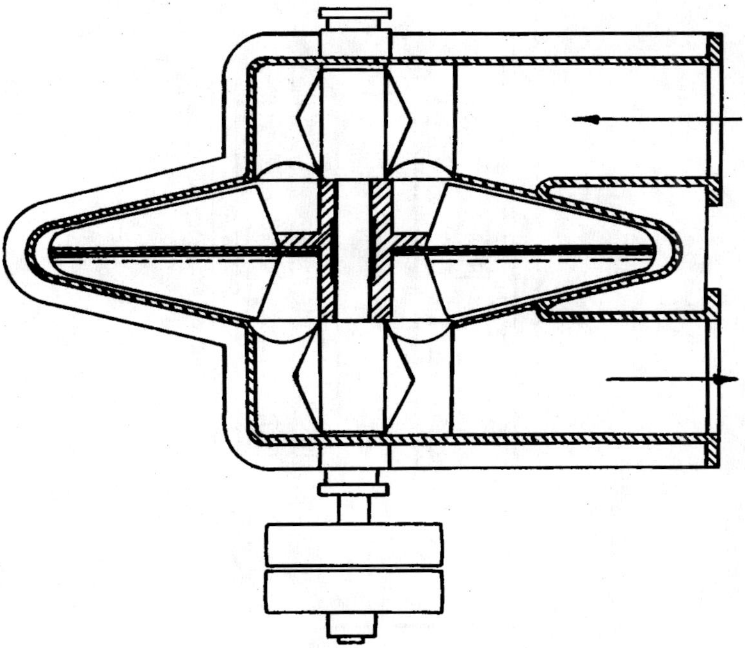

FIG. 11. THE CROSSLEY CENTRIFUGAL TAR EXTRACTOR

THE VOKES FILTER*

Fig. 12 shows a filter that has been developed by Messrs Vokes Ltd in England.

In this filter the gas sucked in by the engine enters a circular chamber at the bottom of the cylindrical filter case. Thence it passes upwards through a ring of tubes, much as water passes along a water-tube boiler, and so reaches the conical mouth of a large chamber. Down this chamber

* From the *Autocar,* 10 November 1939.

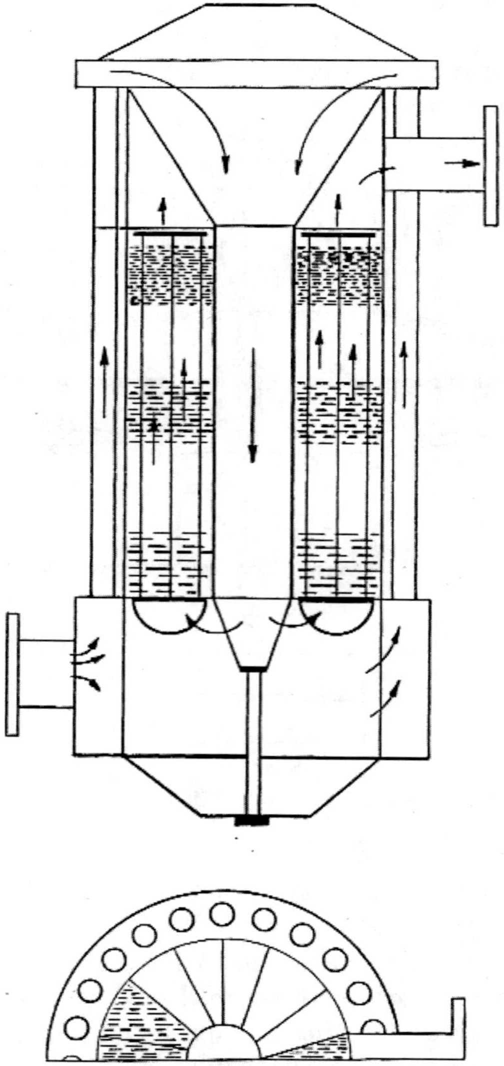

Fig. 12. THE VOKES FILTER

the gas passes, losing many of its impurities in the process, and then turns upwards once more along a series of wedge-shaped passages. In these passages are three sets of brushes looking like the ordinary bristle brushes used for cleaning flues, except that they are all on a common spindle, or rather on three spindles.

The first brush has its bristles set coarsely, in the next the bristles are closer, while the third brush is of much finer quality. Even finer brushes can be added at need. Each set of brushes is circular and is at an interval from its neighbour, and the complete unit is wedge-shaped to fit the container. As the brushes are only held in by the friction of their flexed bristles, the whole can be withdrawn for cleaning quite easily by means of the handle provided, while a bar at the top of the assembly makes it impossible to replace the brushes incorrectly.

Gas, having passed the brushes, passes to the engine from the top of the casing. The deposit is cleaned away by undoing the cover under the filter casing, thus allowing all the dust and solid particles to fall out, and then cleaning the brush compartments with the brushes themselves, then the brushes separately.

The Efficiency of Scrubbers

In the design of a scrubber unit, good cleaning properties are not the sole aim in view. The cleaning must be achieved with as little resistance to the passage of the gas as is possible. For as this resistance increases, so the volumetric efficiency and therefore the power output of the engine decreases.

This resistance, which is called "back pressure," is stated in terms of the pressure drop across the scrubber, and is measured by the column of water the pressure can support.

Where no suction gauge is available, a simple but effective gauge may be constructed by means of a bent tube of piping or a U-tube, preferably of glass, which may be attached to the apparatus at any suitable point. The bend of the U-tube is filled with water and when the plant is in operation the water rises in one side of the tube. The difference in level denotes the back pressure at that point. Readings should be taken on each side of the scrubber to determine the pressure drop caused by the apparatus.

In scrubbers of different producer plants a pressure drop of from 2 to 20 inches at engine speeds approximating to a road speed of 50 miles per hour have been recorded, but it would not seem advisable to exceed a pressure drop of from 7 to 10 inches of water across the scrubber.

Obviously the pressure drop increases with the increased volume of gas drawn at high engine speeds. In the same way the effectiveness of the scrubber decreases with increased gas velocity. For this reason it is advisable that the area through which the gas is drawn for cleaning should be as large as possible without making the equipment too cumbersome.

Considerable difference of opinion seems to exist as to the allowable dust concentration in producer gas plants. While the specifications of the Department of Supply and Development specify an upper limit of 200 milligrammes of dust per cubic metre of gas, a much lower dust concentration seems desirable. One Continental authority quotes 20 milligrammes per cubic metre as the highest permissible dust concentration; but this figure seems difficult to obtain in practice over the wide range of engine speeds and consequent high gas velocities with which the scrubber of a mobile gas producer has to contend.

So far as the writers are aware no tests have been carried out as to the abrasive qualities of the dust carried through to the engine. It would appear that any ash entering the cylinders would be fused at the high temperatures of the explosion and cause cylinder wear. The acidity of the gases would also have an important bearing on this subject, on which much more information is required.

Recent tests give a cylinder wear of 0·374 thousandth of an inch per thousand miles for producer gas, as against 0·250 thousandth of an inch for petrol.

CHAPTER VI

THE COOLER

THE cooling apparatus, or radiator as it is often called, is usually placed in the line before the scrubber. The two principal reasons for the use of a cooler are (i) to reduce the temperature of the gas and (ii) to cause condensation of the water vapour and tar carried in the gas.

Cooling of the gas is necessary for the following reasons:

1. As has already been seen, high temperatures cause damage to certain types of scrubbers and to all those employing filtering cloths.

2. A lower temperature reduces the risk of explosion when the gas is in contact with the air in the mixing valve.

3. The possibility of fire being caused by hot pipes is reduced.

4. The most important reason for the use of a cooler, however, is that it allows a greater weight of combustible mixture to be admitted to the cylinder of the engine, and consequently a greater power output. This is because all gases expand on being heated and contract on being cooled. With the pressure constant, the increase in volume of a gas is $\frac{1}{273}$ of the original volume per degree Centigrade rise in temperature.

This means that a certain weight of gas leaving the generator with a temperature of about 800° C. has a volume approximately three times that of the same weight of gas at a temperature of 30°-40° C., which is the normal inlet temperature of an engine. Therefore, if the gas were not cooled before entering the engine, only a proportion of the weight of charge would be admitted to the cylinders and the power output would be reduced in the same proportion. In other words the volumetric efficiency of the engine would be reduced.

The importance of a cooler varies according to the type of plant and the general arrangement of the installation. If the plant is fitted close to the engine the gas has only to pass through a short line of piping before it is used. In such a case a cooler would be necessary, and if a scrubber containing a filter were being used the cooler would have to be placed between it and the generator.

If, on the other hand, the generator were being placed at the rear of the vehicle, a special cooling arrangement could be eliminated—because the gas has to pass through a long line of piping before reaching the engine and so gets sufficient cooling.

Most coolers are simply a bank of pipes (the number varying according to their size and length) connected by headers; the gas is made to pass through the pipes, and around them the air is allowed to circulate. The pipes should be stood vertically, so that any dust caught in them will fall to a lower surface, thus not hindering the transmission of heat to the atmosphere. In order to avoid passing the gas around a number of sharp corners, and so increasing the resistance of the cooler, the pipes are connected in parallel lines by means of headers (see Fig. 13).

Both these headers should be removable, so that if the pipes become clogged with a mixture of dust and tar it will be a comparatively simple matter to clean them. In addition to this, the bottom header should be equipped with a drain plug for running off the distillate that will be collected in it.

In order to obtain the maximum of cooling from the air, the pipes from which the cooler is made should be of as thin metal as possible, small in diameter, and sufficient in length and number. It is desirable to stagger them, in order to avoid placing them one behind the other. When the units are fitted to the side of a truck the gas is often run up to the back and down again to obtain sufficient cooling.

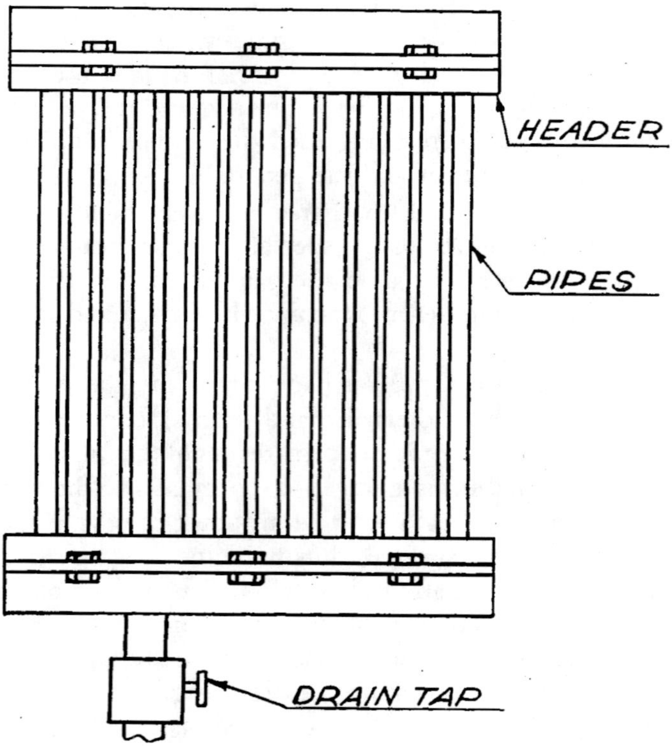

FIG. 13. A TYPICAL COOLER UNIT

USE OF AN ORDINARY CAR RADIATOR FOR COOLING

In England many experimental gas producers have an ordinary car radiator for cooling the gas; but the car radiator, in addition to being a costly unit, is very liable to failure by blocking with dust and tar. This can be avoided to some extent by placing the cooler after the scrubber, but such an arrangement rules out the possibility of using filter cleaners.

Water Cooling

As already stated, the pipes are generally air cooled. But a much more effective unit would be produced if the cooler could be jacketed and a plentiful supply of cold water passed through it. Unfortunately this is not possible in producers for motor vehicles. Wishart Brothers, however, have taken full advantage of a supply of sea water for cooling the gas in the plants they manufacture for motor boats. A small pump is used for circulating the water.

Expansion Chamber

If a gas is made to expand rapidly it cools itself. This is the principle employed in the expansion type of cooler, which can also be used as a dust trap.

The fundamental drawback to this type of cooler is the large size of the chamber necessary to provide sufficient cooling.

CHAPTER VII

FITTING GAS PRODUCERS TO MOTOR VEHICLES

THE great variety of motor vehicles and the varied conditions under which they operate demand that the manner of fitting a gas producer to them may be varied accordingly.

The usual practice, however, is to place the producer on a specially constructed platform at the back of the vehicle in the case of motor cars, and on the side running-boards in the case of trucks.

This platform can either be an all-metal one or have metal stays with a wooden base; but in the latter case it is necessary to protect the wood from burning by a metal covering, or preferably by lagging with asbestos sheeting.

The units comprising the producer are held in position by means of metal bands around the centres, and a number of tie bolts (usually about three) holding them down to the platform.

GAS PRODUCERS FOR MOTOR LORRIES

The principal reason for putting the producer at the side of a truck is that loading and unloading would be handicapped if it were put at the back; in addition to which, the fire and fuel in a producer placed at the back of a truck are often badly thrown around when the vehicle is travelling over rough roads with little or no load. Placing the plant at the back, however, possesses one great advantage in that the gas has to pass through a long line of air-cooled piping before it reaches the engine, and so becomes fairly well cooled. This cooling may even be sufficient to allow the elimination of a special cooling unit. In all but few cases the tray of a motor lorry projects beyond the cabin, and it is into this space that the producer is fitted.

As this space seldom exceeds 15 inches in width, it is often necessary for the plant to project beyond the tray of the lorry. This should be avoided if possible, because there is a likelihood of this projecting part catching in a door or gate, causing the whole plant to be torn from its platform. Thus it may be found necessary to cut away portion of the cabin or tray, or even both; but this also should be avoided.

Whenever possible all the units are placed together, so as to eliminate the danger of leaks occurring in the joints of the pipes connecting them; but the weight of large plants for trucks may be so great that the units have to be split. A not uncommon practice is to place the generator on the left-hand side and the scrubbers and coolers on the right.

Gas Producers for Motor Cars

As already stated, producers for motor cars are usually mounted at the rear. With the older makes of cars the units may be put on the running-boards forward of the front doors. All modern cars have at the back a luggage compartment into which as much of the producer as possible may be fitted, the rest being held on a specially built platform. This platform must be rigid; otherwise, movement may cause leaks in the gas leads.

The French have paid particular attention to the complete enclosure of the producer within the body-work of the car, Messrs Gohin-Poulenc having produced a plant which completely fits in the luggage compartment and even allows closing of the doors. One Australian firm, Messrs

Wishart Bros, have designed a plant to fit into the body-work of a car as shown in the picture of a Ford car (Fig. 14). The more common procedure in France seems to be to fit the plant at the rear and cover it in with a specially constructed streamlined cowling (Fig. 15).

These arrangements detract very little from the appearance of the car; but, apart from the loss of the space normally available for luggage, there is danger of fire owing to insufficient cooling from the atmosphere.

At least one manufacturer is fitting his truck plants in front of the engine. This eliminates the space problem but has several disadvantages. Among these are the damage that would be done to the plant in the case of accidents, the added weight on the front springs, and a danger of the driver's view being obstructed.

The position in front of the engine is a very good one as far as the cooling unit is concerned, since it gets a plentiful supply of cool air, and the very short leads used reduce the possibility of leakage.

All the possibilities cannot be enumerated here; but it is hoped that this general résumé may help designers and fitters in converting motor vehicles to producer gas.

The Use of Trailers

Although only recently adopted in Australia, fitting the producer on a trailer has been generally accepted in France and in England.

This system offers several decided advantages: The greater space available enables grouping of the components; the plant is easily removed for servicing and repairs; and the factory construction of plants complete on a trailer greatly facilitates fitting. Further, those private owners who are interested in producer gas only as a war-time emergency measure may have plants fitted without inter-

Fig. 14. FORD V8 CAR WITH GAS PRODUCER FITTED IN THE BODY-WORK

Exhibited at the Melbourne Motor Show, 1940

Fig. 15. THE GAS PRODUCER FITTED TO THIS CAR PRESENTS QUITE A NORMAL APPEARANCE

ference with the body-work of the car, and so without detracting from their trade-in value. Another point in favour of the trailer is that there is now no necessity to strengthen the rear springs to carry the extra load (2 to 3 cwt.).

Farmers and graziers who may wish to use their producer gas plants occasionally on stationary engines for pumping, driving saw benches etc., will find the trailer method convenient.

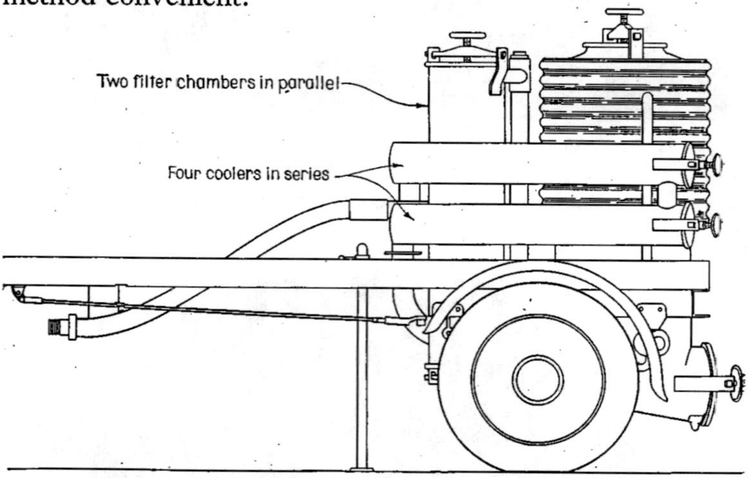

FIG. 16. BRITISH GOVERNMENT'S APPROVED WAR-TIME "EMERGENCY" GAS PRODUCER UNIT

Mounted on approved trailer as recommended by Government Committee. Cross-draught producer with water-cooled tuyere. Water supplied from corrugated iron tank surrounding body of producer. Admission of water or steam is omitted in the interests of simplicity.

By courtesy of *The Automobile Engineer.*

The trailer could be readily disconnected from trucks that have to be loaded and unloaded from the back.

The principal disadvantages of the trailer are that the initial cost is about £20 higher, and on top of this there is a running cost, principally for tyres. Certain difficulties may also be encountered in parking.

The single castor wheel type of trailer is being adopted by a number of makers, in order to overcome the difficulties of backing encountered with the two wheel trailer and also to provide a cheaper unit.

LOCATION OF THE COOLER

Because of its nature and size, the cooler is perhaps the most easily placed unit of a gas producer. In Australia it is usual to put it with the other units; but in France particularly, this arrangement has not been strictly adhered to. In Fig. 17 the cooler is shown strapped to the chassis and running transversally across it, the headers being so placed as to be easily removable for cleaning; this arrangement is suitable only for motor trucks.

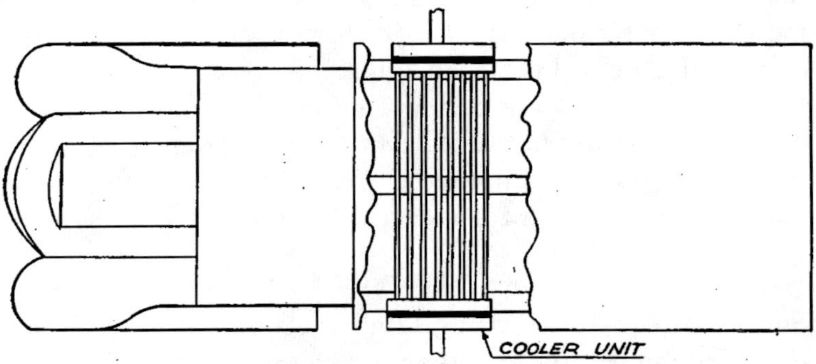

FIG. 17. ARRANGEMENT OF COOLER ON MOTOR TRUCK

One Australian manufacturer is strapping the cooler to the under side of the tray so that it runs lengthwise along it. Although in some cases it is difficult to fit the cooler in this position, it is the most popular and in many ways one of the best arrangements.

Other positions for the cooling unit on a motor truck are behind or on top of the cabin. In the first case there exists a danger of its being damaged by the load carried on the tray, as well as the disadvantage of its being shielded from the air. Although convenient from the point of view of servicing, neither of these two positions is very satisfactory, because the lack of rigidity of the hood of a motor lorry causes leaks in the pipes leading to and from the cooler.

Since it is desirable to have the cooler in contact with the cool air as much as possible, a very good position for it is in front of the radiator; or, for the sake of appearance, between the radiator and the radiator grills. When the cooler is placed in more shielded positions baffles and fins should be provided to pick up the air and circulate it around the pipes.

It is far more difficult to fit the cooler on motor cars than on trucks, because in motor cars it cannot be placed under the body. However, since the all-steel hood of a modern motor car is far more rigid than that of a truck, the cooler can satisfactorily be placed on it. In most modern cars there is a large amount of space under the bonnet where a cooling unit could be placed, but the heat from the engine makes this position unsuitable. It would, however, be a very suitable position for a dust trap or security filter.

Connective Piping

Since it is impossible to build all the units of a producer directly coupled to each other and to the engine, they must be connected by some form of piping, the suitability of which is vitally important to the unit as a whole.

Because of the many bends that have to be made, the first material that would suggest itself would be flexible steel tubing. Experience, however, has shown that this tubing is unable to stand up to the vibration and stresses imposed upon it, and the prospective purchaser of a plant is warned against it, since gas leaks are the *bête noire* of the gas producer operator

Rubber Tubing

As far as flexibility is concerned, rubber hose is equally good, and it will not leak unless the joints are defective. Its disadvantage is that it insulates the gas from the air, and so prevents its being cooled in its passage to the engine.

A combination of seamless steel tubing of about 2 inches in diameter, and rubber tubing for the bends, has proved to be most suitable for this purpose; it does not leak and allows effective cooling of the gas. Steel tubing should be used between generator and cooler, because rubber cannot stand up to the heat of the gases at this stage.

For making the connexion between these materials metal radiator clips are used, care being taken that they do not become loose after a period of use. It will be found advisable to make all connexions between two metal pipes with a bolted double flange joint, a fibre gasket being placed between them.

Drain Plugs

Since all piping has a cooling action, a certain amount of condensation will take place in it and provision must be made for the removal of the condensate.

Towards this end drain plugs should be placed in the pipe line at all low points. These, with some additions, will also provide places where U-tube tests can be made.

Where sagging of rubber hose takes place, allowing the collection of the condensate at the lowest point, provision should be made for its periodical removal; otherwise sudden acceleration will suck the condensate through to the engine, causing spluttering, as well as having a damaging effect on the valves.

To reduce the risk of fire, steel pipes should be covered with asbestos wherever they may come in contact with wood or other inflammable material.

ENGINE FITTINGS

Bearing in mind the previous remarks on the placement of the plant on vehicles, and having chosen the most convenient location for the vehicle concerned, we have to decide what alterations are necessary to the engine manifold and driving controls. It is important that the flow of gas from the producer should be as uninterrupted as possible, and with this end in view short bends or angles should be avoided. It is generally possible to cut away a

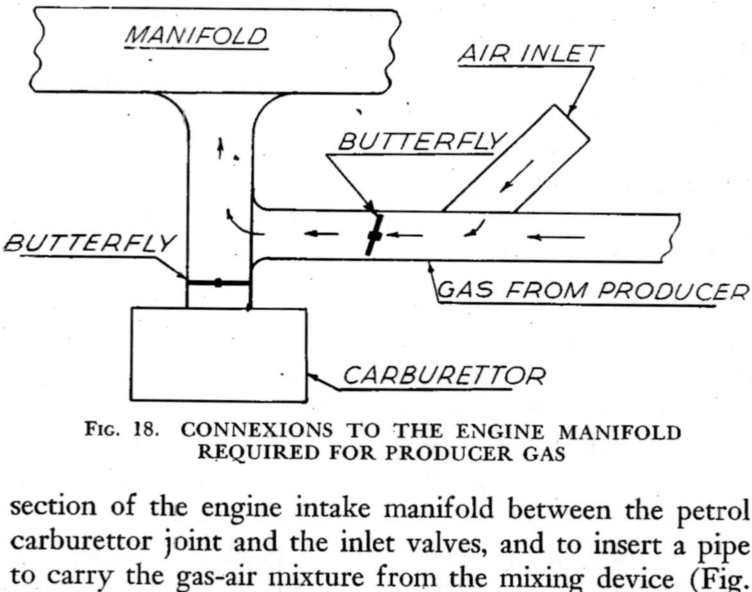

FIG. 18. CONNEXIONS TO THE ENGINE MANIFOLD REQUIRED FOR PRODUCER GAS

section of the engine intake manifold between the petrol carburettor joint and the inlet valves, and to insert a pipe to carry the gas-air mixture from the mixing device (Fig. 18). To this pipe is attached the driving control throttle and air-mixing valve, supplied by the manufacturer of the plant.

In this connexion, it should be remembered that the producer gas and air are mixed in about equal parts. The size of the butterfly valves necessary to control these mixtures is a matter for the designers of the plant; but it is important that the air inlet should be large enough to supply the necessary air, for, apart from the fuel loss entailed, a "rich" mixture will not explode.

This is often a source of trouble to garagemen inexperienced with producers, since the plant starts up on gas quite easily, but immediately "cuts out"—and the more vigorous the draught (or engine suction) applied to the fire, the more impossible it is to secure satisfactory running. The most suitable arrangement is to provide an air-mixing valve for slow and moderate speeds and an extra air device for high speed running (see also Chapter XIII).

Controls

Different makers fit different sets of controls when converting motor vehicles to producer gas. In some cases the petrol controls are used for gas, a hand control being fitted for use with petrol either separately or in conjunction with gas.

The most convenient method is to retain the petrol controls provided and to fit an additional foot accelerator for producer gas as well as an extra air control. Provision should be made for the petrol supply to be turned off completely when long runs are being made on gas; otherwise a certain amount of petrol will be drawn in through the idling jet and, since this adds very little to the power output of the engine, it may be regarded as petrol wasted.

In this system when a stiff climb or heavy pulling requires a change over to petrol, the change can be effected immediately, provided the petrol is turned on; and, since an uphill stretch is almost invariably followed by a similar downhill run, a change back to gas can easily be made. This method is much preferable to that often suggested, namely, using a certain amount of petrol all the time to boost the gas. Our experience has been that when the petrol butterfly valve is opened to admit a certain amount of petrol mixture, the draught of the generator is correspondingly reduced, the fire dying away and the reduced gas supply involving a larger opening of the petrol throttle, which again reduces the draught at the generator, until finally no gas is available and the engine must be run entirely on petrol.

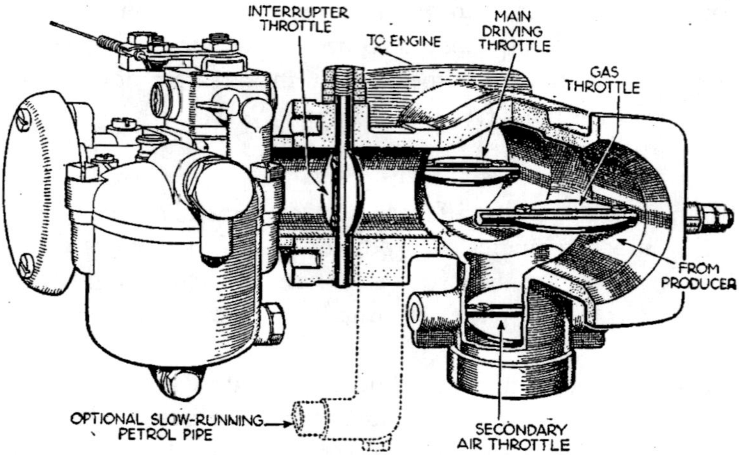

FIG. 19. THE SOLEX GAS-PETROL CARBURETTOR

An English firm, Messrs Solex, has produced a very interesting instrument, illustrated in Fig. 19, for use when an engine is running either on petrol or on producer gas, the two separate types of fuel being available at any time.* The carburettor, as Fig. 19 shows, is made up with an extension to its body so arranged that the carburettor can be cut off entirely from this extension by an interrupter throttle. On this happening, the main throttle from the producer is opened, and thereafter the supply of mixture to the engine is controlled by a third throttle in the ordinary manner.

* From the *Autocar*, 20 October 1939.

In addition, there is also a fourth throttle which admits air at atmospheric pressure, as and when required. With this arrangement the slow-running system may have to be altered considerably, since its orifice should be close to the throttle that governs the supply of the engine intake pipe. A special fitting can be supplied for this purpose. No second throttle pedal is necessary; the ordinary pedal operates the gas throttle and the interrupter throttle, one opening as the other closes, there being a slight overlap so that the change from producer gas to petrol occurs progressively.

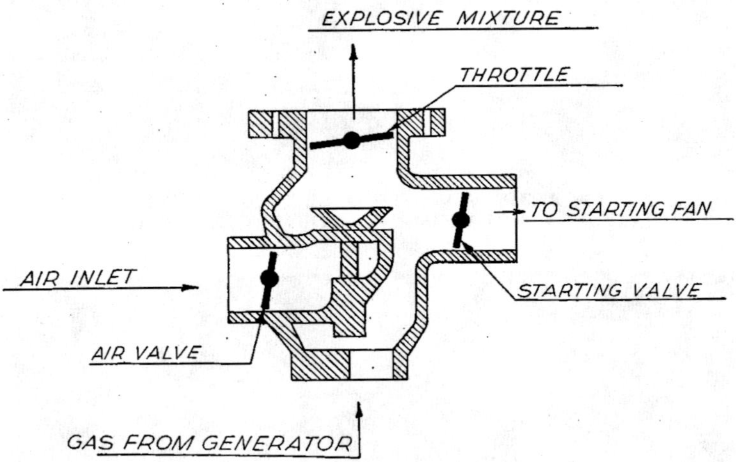

FIG. 20. THE BERLIET MIXING VALVE

In another system the following device is employed: Petrol is used for starting up, and an ingenious valve is fitted to the intake pipe to control the admission of petrol or producer gas mixtures, either separately or together. The valve is of the tapered barrel type, held on its seating by hinge spring clamps, so that jamming is prevented. The barrel is formed with a tapered passage which, in the vertical position, is a continuation of the intake pipe, though it

Fig. 21. AN ARMSTRONG SIDDELEY ENGINE WITH THE NECESSARY ATTACHMENTS FOR PRODUCER GAS

Fig. 22. THIS ENGINE IS EQUIPPED WITH A DOUBLE INDUCTION MANIFOLD, AIR VALVE, AND NORMAL PETROL CARBURETTOR

seals the petrol mixture from the carburettor. One side of this passage, however, is cut away, so that when the barrel is turned to cut off the producer gas, the petrol mixture can be drawn into the engine.

Within the barrel is an ordinary butterfly valve mounted on a spindle, with a lever so connected with the carburettor butterfly valve that both are controlled by the same accelerator pedals. Movements of the barrel do not affect the butterfly valve inside; consequently the producer gas, or petrol mixture, control is independent of the accelerator pedal control, which regulates only the quantity of either mixture supplied to the engine.

Figs 21 and 22 show engines which have been converted to producer gas. The Berliet mixing valve is shown in Fig. 20.

Wishart Patent Gas and Air Mixing Valve

The Wishart mobile type mixing valve is made in two sizes (measured at the throttle diameter opening) of 1½ and 2 inches, and Fig. 23 shows the adjustments. Turning the knurled adjusting screw (2) to the left puts more tension on the spring and tends to make the gas and air mixture richer, while turning to the right makes the mixture leaner. This adjustment is set in the best position for easy starting.

The running adjustment is made by setting the air inlet to regulate the amount of air entering the mixing valve, and is made while the engine is on load at its normal speed.

Idling

In addition to the flat spring loaded valve with an adjustment by a knurled screw, there is a small gas passage connecting directly with the induction passage to the engine below the throttle, and this is regulated by a screw (3), allowing a rich idling mixture to be drawn in by the engine.

Turning this idling screw to the right decreases the supply of gas and makes the mixture leaner, while turning to the left makes the idling mixture richer.

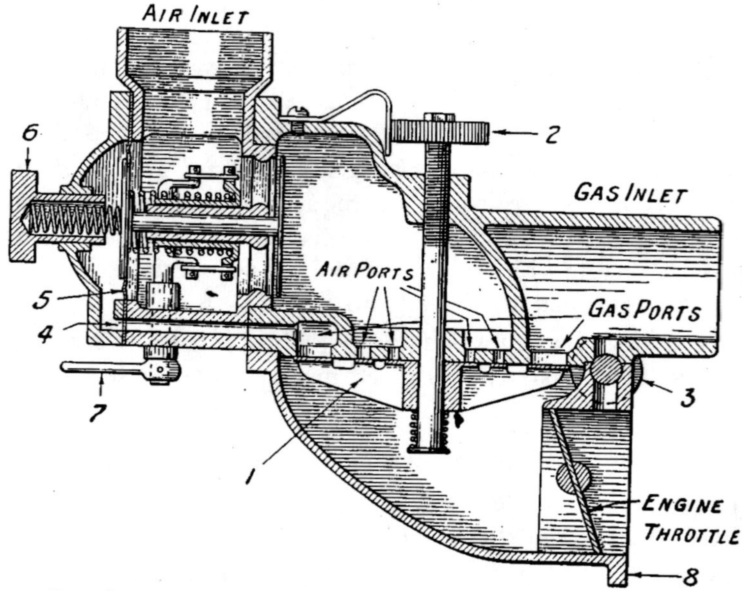

FIG. 23. THE WISHART GAS AND AIR-MIXING VALVE

1. Triple disk valve controlling flow of gas and air. 2. Knurled adjusting screw for varying spring pressure on the disk valve. 3. Idling gas adjustment screw. 4. Gas passage to rear of diaphragm. 5. Diaphragm operating the air valve to equalize pressure of air and gas. 6. Adjustment for outer spring of diaphragm. 7. Combined diaphragm inner spring adjustment and air choke. 8. Flange for bolting to the engine inlet manifold.

Air Choke Button

The air inlet is controlled by a diaphragm operated by the suction pressure of the gas, to which a lever (7), designed to be coupled to a choke button control on the dash, is fitted. The movement of the choke button as it is pulled out will operate this lever (7) and make the mixture richer and richer till, when the button is right out, the air is shut off altogether. To start the running adjustment on this air control, push the gas choke button right in, and while the car is on the road adjust the knurled screw (6) on the air inlet end of the mixing valve to the best position. Screwing in makes the proportion of air greater and consequently weakens the running mixture to the engine.

Wishart mixing valves are made so that the various parts have alternative positions, to fit different engines. But when any part is reversed, carefully note whether any passage which may influence the operation of that part is covered over in the new position.

For engines which require an air cleaner, the original air cleaner from the carburettor can be fitted to the mixing valve air intake, which is designed to be readily machined to fit it. When the carburettor is retained, a second air cleaner may be fitted, or a branch pipe connected to both the mixing valve and carburettor will enable the one air cleaner to serve both.

CHAPTER VIII

MEANS OF OVERCOMING POWER LOSS

As stated in the chapter on the theory of producer gas, the high proportion of non-combustible gases (mostly inert nitrogen) in the producer gas mixture lowers the specific power output of an engine as compared to petrol. This in an engine of $4\frac{1}{2}:1$ compression ratio is approximately 40 per cent, as shown in the accompanying graph (Fig. 24).

This is undoubtedly a very serious power loss and, in the case of tractors, the most serious impediment to the conversion to producer gas—tractors being in general worked nearer to full capacity than motor vehicles.

It is necessary to consider by what means this deficiency can be minimized. The following are some of the means suggested: (i) raising the compression ratio, (ii) fitting a new engine, (iii) boosting with petrol, (iv) supercharging, (v) altering the gear ratio.

Raising the Compression Ratio

As shown in the graph (Fig. 24) there is a 40 per cent loss of power at $4\frac{1}{2}:1$ compression ratio. Raising the compression ratio to 6:1 reduces this loss to approximately 22 per cent.*

Other authorities give rather different results. Goldman and Jones† quote Beale as stating that a gain of 8 per cent

* Rennie, E. J. C., "The Application of Producer Gas to Motor Vehicles," *Journal of the Institution of Engineers, Australia,* 1930, p. 101.

† *Journal of the Institute of Fuel,* February 1939.

is obtained by raising the compression ratio from 5:1 to 6:1; 6 per cent from 6:1 to 7:1; and only 4 per cent from 7:1 to 8:1.

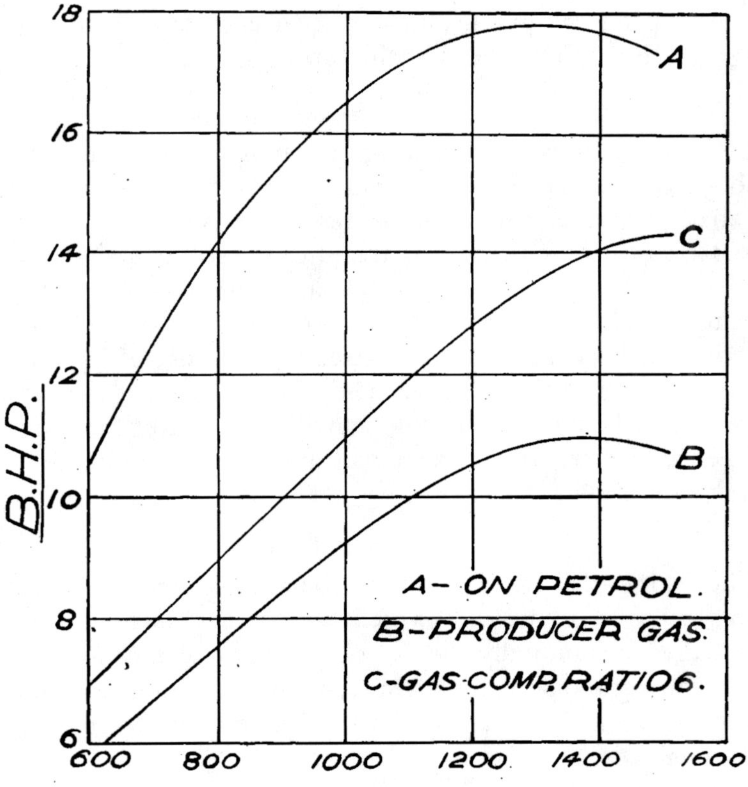

FIG. 24. GRAPH SHOWING THE POWER LOSS OF PRODUCER GAS AS AGAINST PETROL

The compression ratio can be increased by having plates attached to the piston heads or by having new pistons fitted, with an increased height above the gudgeon pin. The latter is the best means of obtaining the required result, plates being difficult to attach without risk of loosening, owing to expansion and contraction and to the fact that the kinetic energy of the plates tends to tear them off their studs at the top of each stroke. Needless to say, a loose plate can practically wreck an engine by jamming against the cylinder head at the top of the stroke.

Tolerances Required for Producer Gas

In fitting pistons for producer gas it is advisable to use about 30 per cent greater clearance than when fitting for petrol. This is necessary because of a slight tackiness imparted to the oil from the traces of tar still contained in the gas. This also applies in the case of new valve guides being fitted, inadequate tolerances causing sticking of the valves with consequent poor running and loss of power, especially when the engine is cold. It is of interest to note that a fairly well worn engine will give better running when converted to gas than will a new one of the same make and model.

Fitting a Larger Engine

The fitting of a new engine has been suggested by some writers as a means by which the power loss can be remedied. Admittedly this is a simple and effective way of overcoming the power loss, but economically impracticable in most cases. One may well ask: Where may the engines be obtained and what will they cost?

Increasing Engine Capacity

When engines are fitted with removable cylinder sleeves, as is the case in a number of tractors and some truck engines, increased cylinder capacity may be obtained by fitting over-size cylinder sleeves and new pistons. In this way it is claimed that up to 12½ per cent increase in cylinder capacity, and hence similar power increase, may be obtained. At the same time increased compression ratio may be obtained and, when it is possible, this seems a practical method of overcoming power loss without unduly high conversion costs.

This is particularly applicable when the engine to be converted has reached the stage of wear at which a re-bore is necessary.

Boosting With Petrol

It has been suggested that more power may be obtained by running on a mixture of petrol and producer gas. This is certainly possible; but it requires a high degree of intelligence as well as considerable experience to adopt this method with the fitments usually provided by the manufacturers of producer gas plants.

The reader is referred to the section on engine fittings (p. 80), where it is pointed out that opening the petrol throttle reduces the air suction in the generator and allows the fire to die away. If the opening of the petrol throttle is persisted in, particularly during periods of light running, the furnace temperature falls to a point at which the generator ceases to function.

However, with a properly fitted plant, an experienced operator can make use of petrol to boost the power in hill climbing and heavy pulling. In the hands of an unskilled or careless operator the practice is open to considerable abuse.

Supercharging

If advantage is taken of the anti-knock value of producer gas, a supercharger may be fitted with an appreciable gain in power. According to Goldman and Jones,* the following result is obtained:

In the case of a 6:1 compression ratio engine, super-

* *Journal of the Institute of Fuel,* February 1939.

charged by 4 lb. per square inch above atmospheric pressure, the gain in power is 36 per cent, not taking into account the power used to drive the supercharger. In contrast to this, raising the compression ratio from 6:1 to 7:1 without supercharging only shows a gain of 6 per cent. Moreover, in the one case the maximum pressure rises by 120-150 lb. per square inch for 6 per cent increase in power, whereas in the other it rises by only 260 lb. per square inch for an increase of 36 per cent in power.

There exist two methods of fitting a supercharger. First between the carburettor and the engine, and secondly with the carburettor between the supercharger and the engine. The first method is the one usually adopted; so for producer gas, the gas would be admitted between the carburettor and the supercharger.

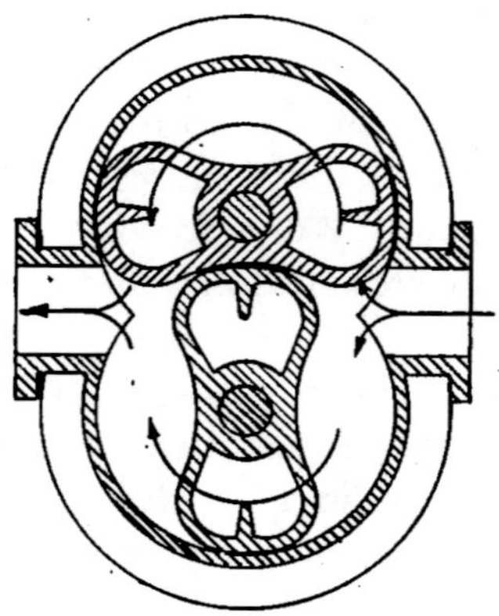

FIG. 25. THE ROOTS BLOWER TYPE OF SUPERCHARGER

There are two main types of superchargers, the centrifugal fan type operated at speeds approaching 40,000 r.p.m., and the twin rotor or vane type working at about one and a half times engine speed.

The Roots Blower

The Roots blower is the most commonly used supercharger. It consists (Fig. 25) of two rotors shaped not unlike the figure eight. These rotors turn in opposite directions and act as a pump in forcing the gas into the cylinders by way of the engine manifold. The advantage of the Roots blower is the low speed at which it operates. The centrifugal type is efficient only at high constant speeds, such as those at which aircraft operate.

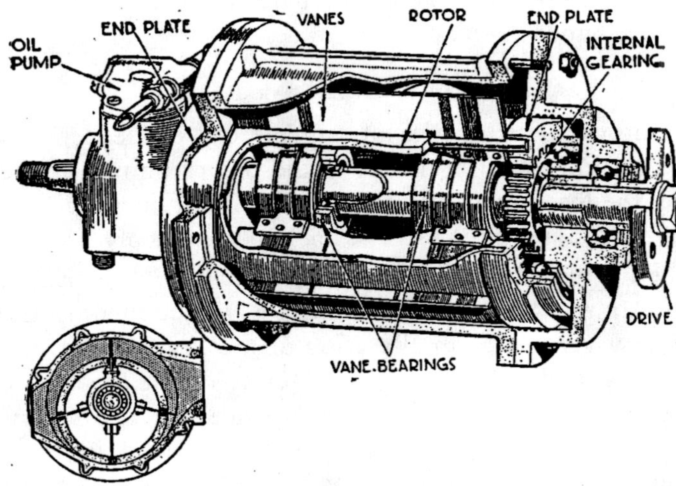

FIG. 26. THE CENTRIC VANE TYPE OF SUPERCHARGER

Fig. 26 shows the centric vane type of blower, which is distinguished by its remarkable freedom from noise.

To restore fully normal speed power and acceleration involves the use of a supercharger capable of giving approximately 7.5 lb. per square inch, this being well within the range of any modern blower such as the Roots.

It is of interest to note that one French manufacturer, the S.U.C.A., fits a supercharger as standard practice. A combination of supercharging and raising the compression ratio will give a power in excess of that of petrol.

At the present time a supercharging unit is just as difficult to obtain as a new engine, but were it necessary, and the demand sufficiently great, the manufacture of such a unit should not be beyond the capabilities of Australian industry. This method would seem particularly attractive in the conversion of tractors to producer gas, the high power output being of such vital importance and the first cost of the supercharging unit being offset by the savings effected with continuous running.

Alteration of the Gear Ratio

The top gear performance of any motor vehicle depends on the relations existing between the power output of the engine, the loaded weight of the vehicle, and the gear ratio of the crown and pinion in the back axle assembly.

When a manufacturer fitted the same model engine in both touring car and 30 cwt. truck, as was commonly the case a few years ago, an alteration of the gear ratio in the back axle assembly was the means used to enable the engine to handle the extra load. In consequence, the revolutions per minute of the truck engine were increased in comparison to the road speed of the vehicle. It is a well-known fact that the most suitable top gear speed of most modern trucks is higher than the maximum loaded speed allowed by the transport regulations of most Australian States. Hence, a reduced road speed as compared to engine revolutions would be no disadvantage even on petrol, and would enable the engine to handle the load on producer gas without the undue labouring and loss of speed so detrimental to the furnace condition of a gas producer.

There are in Australia a number of firms manufacturing

spare parts for the motor car industry, and no difficulty should be experienced in obtaining the necessary parts. While the cost would vary with different vehicles it should not be more than £15. In determining a suitable gear ratio for producer gas, consideration must be given to the petrol top gear performance of the vehicle to be fitted. It would not be advisable to seek to achieve the same top gear performance on producer gas, since, theoretically, this would necessitate an increase in the gear ratio proportionate to the deficiency in specific power output of the engine on gas.

In practice this is not the case, for the slow-burning properties of producer gas enable the engine to "hang on" in top gear, even while losing revolutions, without adversely affecting the engine. It is advisable, therefore, to choose a gear ratio that will provide a reasonably good top gear performance on gas without reducing the maximum top speed of the vehicle.

Just what this would be depends, as stated previously, on the particular performance of the vehicle to be fitted, and a considerable amount of experimental data would need to be compiled before a general specification could be evolved. It would appear, however, that a compromise of about half of the theoretically necessary increase in gear ratio should be made. In other words, where a power loss of 40 per cent is sustained with producer gas, a 20 per cent increase in the gear ratio should suffice. Taking into consideration both the mechanical and economic aspects of the problem, our opinion is that, while modern cars would need no alteration for fitting with producer gas, an alteration of the back axle ratio is desirable in trucks, and for the fitment of tractors a supercharger is possibly the best means of overcoming the power loss.

CHAPTER IX

DRIVING OF VEHICLES ON PRODUCER GAS

WHILE at first sight the extra controls are somewhat bewildering in vehicles driven by producer gas, people used to driving on petrol will experience little difficulty. The number of these controls and the means of operating them vary with the different plants, and instructions will have to be obtained from the manufacturers. Apart from the manipulation of these controls, however, there are certain other points that must be observed by the good gas producer driver.

From previous chapters it has been seen that the production of a good gas is dependent on a hot fire, the temperature of which is governed by the engine suction. At low speeds this suction is small, and thus the fire begins to die out, a poorer gas is produced, and the engine runs badly. To bring its performance back to normal necessitates raising the engine speed, and in order to maintain the same rate of travelling a lower gear will have to be used.

When approaching a hill it will be found advisable to change gear, in order that a plentiful supply of gas will be available when needed; and if, even then, the gradient cannot be taken on gas, a change over to petrol will be necessary. When vehicles using producer gas are to be continuously operated in hilly country, supercharging, fitting of a larger engine, or alteration of the back axle gear ratio will be necessary.

Changing Over From Petrol to Producer Gas

Perhaps one of the trickiest operations to master on a gas producer vehicle is the change over from petrol to gas. This will now be considered in detail. After making sure the generator is clean and fully charged with fuel and the lid tightened down, light the generator by placing a piece of kerosene-soaked waste in contact with the fuel, applying a match and starting the engine on petrol. The engine is run a little above idling speed with the gas throttle well open, so as to create a forced draught which soon lights up the fire. The engine is left running in this condition for about 5 to 10 minutes, depending on the fuel and type of plant.

It should be remembered that in the early stages of lighting little or no combustible gas is formed in the fire; hence the gas from the generator will not explode and too much must not be admitted to the engine or it will cut out. At this stage the choke of the petrol system must be judiciously used, the engine drawing part of its air supply together with a proportion of non-explosive gas through the generator, and still more air through the mixing valve.

Later, when an explosive (or more correctly a combustible) gas is formed in the furnace, a richer gas will be available and the engine will be getting a mixture of producer gas, petrol vapour and air. The choke may then be released and the petrol butterfly valve gradually closed to increase the proportion of producer gas admitted to the engine.

Should the engine fail to take off on gas, intermittent bursts of petrol and gas running at high engine speeds should build up the fire to a stage where it is possible to take off on gas.

It should also be remembered that until the generator has been alight for some little time the supply of gas available for sudden acceleration or heavy pulling will not be sufficient, and there will be a danger of the engine cutting out. However, the driver will soon become accustomed to the feel of the engine on gas, and his judgment will then prevent minor mishaps of this description.

Regulating the Supply and Mixture of Gas

The driver must realize that he is both engine driver and stoker, and has to drive in such a manner as always to provide sufficient gas for the immediate requirements of his engine. The fire cannot suddenly be built up to provide gas for an emergency, and this lack of flexibility must be accepted as inherent in a gas producer. Moreover, the fire, having once been built up, will not suddenly die down and cease to produce gas just because road conditions do not require its use. This is made noticeable where traffic conditions or other causes necessitate a short interval of idling during a long run. As soon as the load is taken up and full power is required, an over-rich mixture will be produced, and the extra air will have to be used to counteract the slight pressure which has been built up in the generator and scrubbers. This is rather puzzling when first experienced.

When an extra air control is provided it must be used judiciously, for gas will not stand the same variation of control as petrol will. An over-rich mixture will not fire, and too lean a mixture reduces the suction in the generator and allows the fire to die away. For this reason, too much air must not be used on short runs with intermittent stops. On long runs, however, at moderate or high speeds the air control should be opened up to where the greatest power is obtained.

Considerable difference in the behaviour of the gas producer plant is noticed with different species of charcoal. Mallee charcoal is highly reactive and burns freely, thus keeping a good furnace condition even during idling or with intermittent loads. Charcoal from timber grown on wet soil or in high rainfall country is much slower burning and requires more draught to keep it alight, and the extra air must be used sparingly in order to maintain good furnace draught.

On long downhill runs the furnace condition can be maintained by switching off the ignition and opening the gas throttle (or accelerator). This should be done near the bottom of the hill to ensure a plentiful supply of gas to climb the next hill. No harmful result to the engine will be experienced, as would be the case with petrol.

CHAPTER X

THE MAINTENANCE OF GAS PRODUCERS

ONE of the first objections raised by the critics of gas producers is that their maintenance demands much more time than that of petrol-driven vehicles. However, it is essential that it be carried out regularly and efficiently as prescribed by the manufacturers of the plant.

WATER CLEANERS

If a primary water cleaner is used, drain the water section every 100 miles, scrape out the ash and fine charcoal, and refill to the correct level with clean water. It is necessary to maintain the correct water level at all times during running, for if the water level falls below the gas spreader, gas will escape without being cleaned.

OIL CLEANERS

When an oil cleaner of the impingement type is used, see that the level of the oil does not fall below that indicated by the manufacturer. This seldom happens when a primary water cleaner is also used; in fact, the opposite is usually the case, since condensation of the water vapour in the gas causes the oil level to rise by an addition of water. This is not detrimental, for the oil retains its cleaning properties even with a high proportion of water. After being in use for some time the oil takes on a greenish hue and becomes thick and granular; it should then be changed and fresh oil added according to the manufacturer's instructions.

Dry Cleaners

Remove the accumulations of dust from the ash box or cyclone cleaners daily or about every 100 miles of travel. Inspect the cleaning medium (wood, wool, sisal, horse-hair, felt etc.) to see if it is stained by impurities or dust. Stained layers should be removed and washed with petrol, or replaced by new material.

Oil-soaked pads of sisal etc. should be cleaned every 500 miles, and after cleaning resoaked with oil and replaced.

Filters

Filters should be removed according to the maker's instructions and dirt shaken from them. Owing to the varying types of filters used it is impossible to give a general instruction, but it is imperative that they should be attended frequently, since clogged filters will impede the gas flow and cause loss of power, as well as allowing dust to pass to the engine.

The Generator

When a grate is used it should be cleaned every morning. A clogged grate is one of the most frequent causes of power loss in up-draught generators. Clinker should be removed daily before lighting the generator. When clean charcoal is used distances of 500 miles or more may be travelled without cleaning the grate or tuyere. With dirty charcoal the writer has experienced a badly clogged grate after 20 miles of running. (It cannot be too strongly emphasized that dirty charcoal is the most frequent cause of furnace trouble, and for this reason only charcoal burned in a kiln or steel retort should be used in a gas producer.) Whenever this occurs the generator should be completely emptied before commencing another day's run.

Dirty charcoal may be freed from dirt by washing in water and spreading out to dry in the sun.

THE COOLER

It is important that cooler tubes should be kept clean both internally and externally, since a coating of soot prevents free transmission of heat and adversely effects the efficiency of the cooler. Clean the cooler regularly according to the manufacturer's instructions.

DETECTING FAULTS

The Engine Loses Power

Cause	*Remedy*
1. The fuel is too low.	Add more charcoal.
2. The fuel is too wet.	This will improve as the generator becomes hotter.
3. The grate or tuyere is clogged.	Clean with cleaning tool.
4. The cleaner or filter is clogged (this will be denoted by increased back pressure, preventing full use of air).	Clean material in cleaner or filter and replace.
5. The gas outlet is clogged.	Inspect piping at bends for pieces of charcoal.

Explosions in Scrubbers

Explosions in scrubbers are caused by inlet valves sticking. An engine overhaul is necessary. Also inspect for air leaks in the gas line.

Leaks in Generator or Piping

Leaks in generator or piping may be detected by lighting some damp straw or paper in the empty generator and closing the lid; the slight pressure generated will force the smoke through any leaks.

Every three months the piping and tubing carrying the gas from the generator should be inspected for clogging, which is caused by condensation of the tars contained in the gas. This also applies to the engine manifold and gas butterfly tube. Clean these by washing with kerosene or petrol and brushing with a stiff bristle brush.

CHAPTER XI

FUTURE DEVELOPMENTS OF THE GAS PRODUCER

NONE of us is able to foretell exactly the future development of the gas producer and how far off its perfection may be. That this will be controlled more by economic conditions than by developments of the motor car is certain, but we can only speculate on the future economic condition of the world. We can, however, make an intelligent guess at the lines on which the gas producer will develop.

Thousands of plants have been in operation for a considerable number of years in France, Italy, Germany, Australia, England, Russia, and Japan, so that it is unlikely that any radical changes will be made in producer design; instead there will be a gradual perfection of the systems now employed.

The gas producer as we know it to-day has three fundamental disadvantages which have to be overcome. These are: (i) a loss of power as against petrol, (ii) a large and dirty apparatus, and (iii) the length of time required for starting.

OVERCOMING POWER LOSS

In Chapter VIII ways of overcoming the power loss have been considered, and just which of the means mentioned will be adopted in Australia only time can tell.

If there is sufficient demand, the motor companies may find themselves forced to fit larger engines to the vehicles they manufacture, those already in operation having compression ratio and/or gear ratios changed.

It is also possible that some means of boosting with petrol may be developed; but the petrol used would be expensive and the advantage of the gas producer system offset to a certain extent.

Of course, if it were possible to avoid the double dilution of the gas with nitrogen by extracting this inert element from the air before it enters the generator, or from the gas after it leaves the generator, none of the above means would have to be resorted to, and producer gas would compare favourably with petrol.

Appearance and Size of Plant

The public demands that the modern motor car should be a thing of beauty, and anything that tends to detract from this beauty will naturally meet with severe criticism. Therefore, the appearance of the gas producer will have to be considerably improved if it is to meet with public approval. This end may best be arrived at by decreasing the size of the plant, and making it in a shape that will fit into the body-work of the car, or in a shape that can be covered with a streamlined cowling.

A motor car engine is certainly not beautiful, except perhaps to the enthusiastic engineer; but when cowled in it presents a pleasing appearance which increases its selling value. It is natural to expect that within a short time gas producers will be similarly disguised.

It is also probable that trailers will become more popular; already a number of ultra-modern ones are in operation.

It is quite feasible, too, that the size of the plant will be considerably reduced in the future. For instance, improvements in scrubber design, or the discovery of better cleaning agents, will reduce the size of the scrubber.

Fuels

An organized distribution of briquetted fuels, making them available to the motorist at almost any place, seems to be the only solution to the problem of reducing generator size. It would then be unnecessary to provide for a long range of travel without refilling.

With ordinary charcoal the range of travel for one filling is approximately 100 miles, although this depends on the size of the generator; but with a briquetted fuel such as carbonite (see Chapter III) either four times the distance may be travelled with one filling, or one-quarter the size of generator may be employed.

In conclusion, it may be stated that a briquetted fuel, because of its greater compactness and lack of objectionable "fines," presents the most feasible means of reducing both the size and dirtiness of a gas producer.

It will be difficult to reduce the cross-sectional area of the generator much below what it is to-day and thus any decrease in capacity will have to be brought about by a reduction in the height. This is because it is necessary, in order to prevent the walls of the generator from burning out, to have a lining of cooler charcoal between the walls and the fire zone. The necessary space for this has already been reduced to a minimum.

With regard to fuels, it is of vital importance that the supply should be organized and standardized throughout Australia. For this, Government control and assistance seem necessary.

An officer of the Council for Scientific and Industrial Research has prepared provisional specifications of charcoal for gas producers (see Appendix II). Legislation should be enacted to ensure that charcoal sold for this purpose conforms to these specifications.

It is particularly important that partly burned charcoal of a high volatile content should not be used, since the tarry substances cause gumming, and the acids corrosion, of an internal combustion engine.

Starting Time

With the modern petrol-driven vehicle it is possible to start and be on a road in less than a minute. Although this may never be possible with producer gas, at least the 5-10 minutes required at present will be considerably reduced.

Before a generator can produce a good gas it is necessary to heat a large quantity of fuel to the required temperature. But if a means of restricting the size of the fire zone were devised it would be possible to generate a good gas in a short space of time, make a change over to gas, and then allow the size of the fire zone to be increased so as to generate the volume of gas required. Panhard has attempted this by introducing a second tuyere which is used only for starting purposes.

In addition to the major developments pointed out there will be, in all probability, many minor ones. Among these might be the marketing of charcoal in paper bags that can be dropped completely into the generator. Special refilling apparatus will also be developed as the demand warrants.

Banking

The banking of the fire at night is quite a feasible method of getting a quick start in the morning. This is achieved by lifting the lid and leaving it sufficiently loose to let a very small amount of air pass through the generator. As little as 3 or 4 lb. of charcoal consumed can keep a generator alight for 12 hours or so; future plants may incorporate means of supplying air for this purpose, as well as safety devices to deal with fire and poisoning risks.

In conclusion, the plant of the future should tend to become: (i) smaller and neater, (ii) quicker to start, (iii) safer from fire risk. In addition to this it should have (iv) means for automatically and more accurately controlling the admission of water or steam to the generator according to the variation in power output and the temperature of the furnace, and (v) increased range of travel without servicing.

CHAPTER XII

FRENCH PRODUCER GAS PLANTS

The Panhard Generator*

Fig. 27 shows the Panhard generator, which is of the down-draught type.

It consists essentially of a hopper (A) to which has been fitted an air-tight lid to allow of admitting the fuel. This hopper in turn is formed by an elongation of the cylindrical body (E) which is closed at the lower end by an ash hopper (M). A refractory lining (H) is provided, a steel cylinder (F) being placed outside this. The lining rests on a plate which is riveted to the cylinder. A bell-shaped grate (I) mounted on a pivot can be oscillated by means of a lever (J), thus causing the ashes to fall down into the hopper, from which they can be removed by means of a handhole (N). A second handhole (L) gives access to the combustion chamber. The lower portion of the hopper is made in the form of a truncated cone so as to deflect the charcoal into the centre of the combustion zone which, as previously stated, has a refractory lining. A veilleuse, or vent, is fitted to enable sufficient draught to be created to keep the fire alight when the vehicle is stopped. This vent is closed during travelling. A fan (O) is provided to light the generator.

Method of Operation

The suction of the engine causes air to be drawn through O, whence it passes between the double walls formed by

*Major-General T. R. Williams (Department of Information brochure).

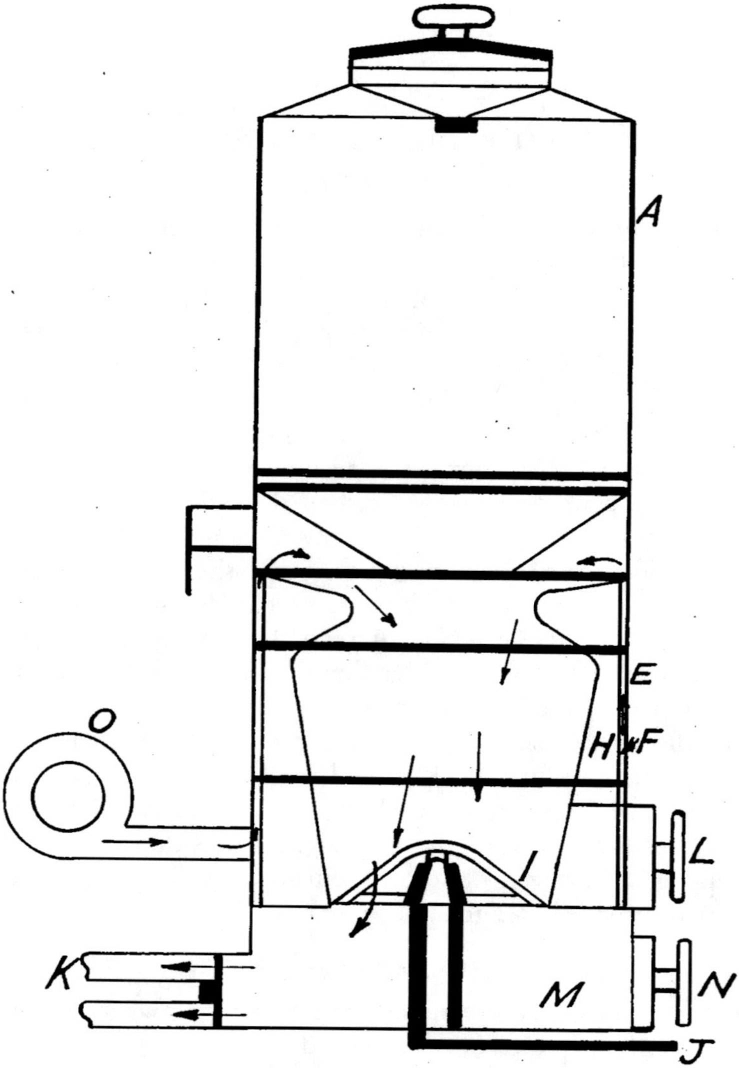

FIG. 27. THE PANHARD GENERATOR

the cylinder (F) and the outer envelope, and is preheated in so doing. It then passes into the fuel where the gas is produced, through the grate and ash hopper, and then to the scrubber through the pipes.

The Scrubber

The Panhard scrubber, which may be classed as belonging to the filter type, has incorporated in it a security filter. It is constructed of a cylindrical body, at the bottom of which is an ash hopper into which the gases from the cooler pass. From here they travel up through a bed of coke, then through a block of filters inside a cloth envelope, and finally through the security filter and on to the engine.

This security filter is composed of gauze with 0·040 millimetre square holes.

The Gohin-Poulenc Generator

The Gohin-Poulenc generator (Fig. 28) is a simply designed cross-draught generator and can burn a variety of fuels.

It is made up of a cylindrical body, of which the lower portion forms the combustion chamber and the upper portion the fuel hopper. A water-cooled tuyere (K) carries the air through the pipe (L) and in to the combustion chamber. The reason for using this pipe is to prevent a jet of flame being expelled if a back fire occurs.

For the purpose of keeping the ash distinctly separate from the fuel (a procedure not always adopted in cross-draught generators) a grate (B) is placed between the combustion chamber and the ash hopper (D), which may be removed for cleaning.

The Scrubber

The Gohin-Poulenc scrubber is in many ways similar to the Panhard, and is of the filter type.

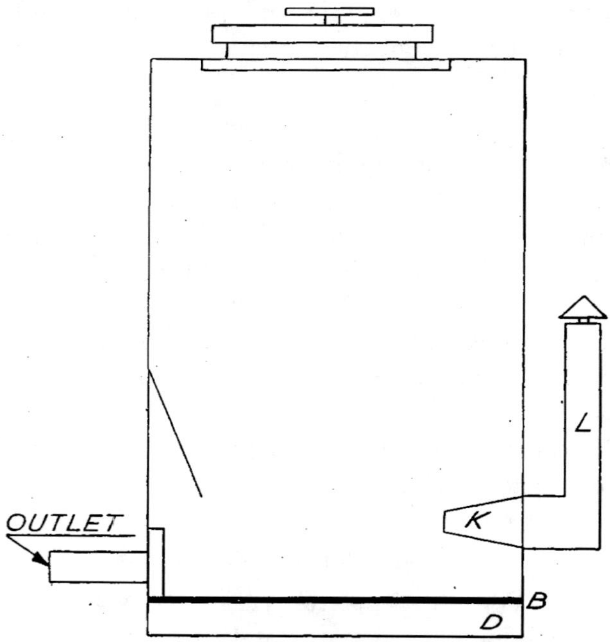

FIG. 28. THE GOHIN-POULENC GENERATOR

The gas is passed through a filtering medium of either pulverized charcoal or a vegetable powder bearing the name of *poudre de liège*. The gas is also sent through a block of filter elements similar to those in the Panhard. Each element is made up of a metallic centre covered with fine cotton cloth and is arranged in such a manner that the dust deposited on the cloth falls to a lower surface, from which it can be removed by a handhole.

A security filter is also placed in the gas circuit and is arranged so that the burning of filter cloths by back firing of the engine is prevented.

The de Dion Bouter Generator

The de Dion Bouter generator is of a rather complicated form (Fig. 29), designed to burn wood—particularly of the green and resinous types. It takes the form of an oval or cylindrical container, the upper portion of which forms the hopper for the fuel. Down the centre is fitted a metal tube which is filled with charcoal and known as a reduction tube. Packed around this tube is the wood, which gradually falls downwards to a common zone, just above

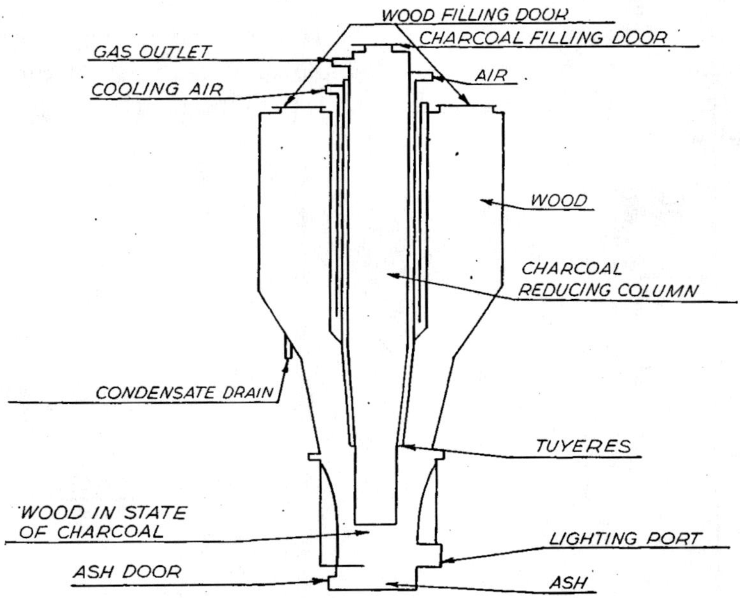

Fig. 29. THE DE DION BOUTER GENERATOR

the grate, where the wood and charcoal mix. Below the grate is an ash hopper. Near the centre of the hopper is fitted a collector, used to drain off the condensate which forms on the internal walls of the hopper.

A number of tuyeres carry the air to the combustion

zone after it has first circulated in the annular space between the concentric tubes, where it is preheated. This arrangement also cools the upper part of the hopper. The gas outlet is placed near the top of the column of charcoal.

The Berliet Generator

The Berliet generator, which is designed to operate exclusively on wood, is of the down-draught type, consisting of a body and a hopper. It has a number of radial tuyeres which connects with the combustion chamber, the air being admitted through a flap valve. The hopper is formed by an elongation of the furnace and is closed by a lid. Fig. 30 shows the principle of operation. Berliet uses a special nickel-steel alloy for all parts which have to withstand high temperatures.

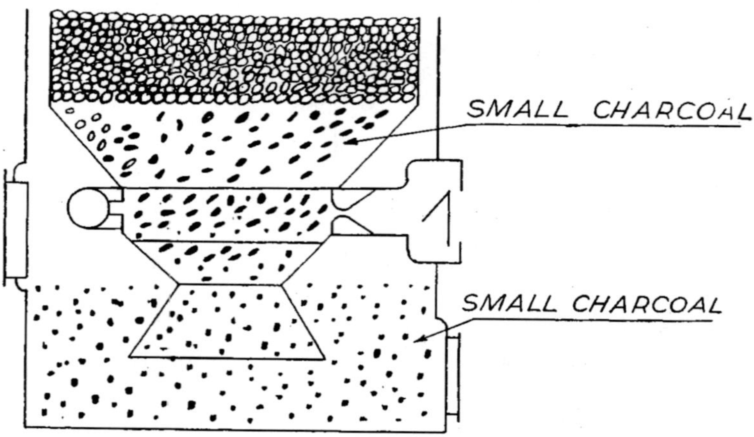

Fig. 30. THE PRINCIPLE OF OPERATION OF THE BERLIET GENERATOR

The Roux Generator

The French maker, Roux, has designed a novel producer, which in appearance is very like a wood stove. It consists of a narrow box 2 feet 6 inches high, 3 feet long, and 12 inches wide, placed on the running-board and designed to use briquetted charcoal ("carbonite"). As Fig. 31 shows, the fuel is fed through two lids on the top of the generator into a hopper leading to the furnace. The grate consists of narrow parallel bars of cast iron.

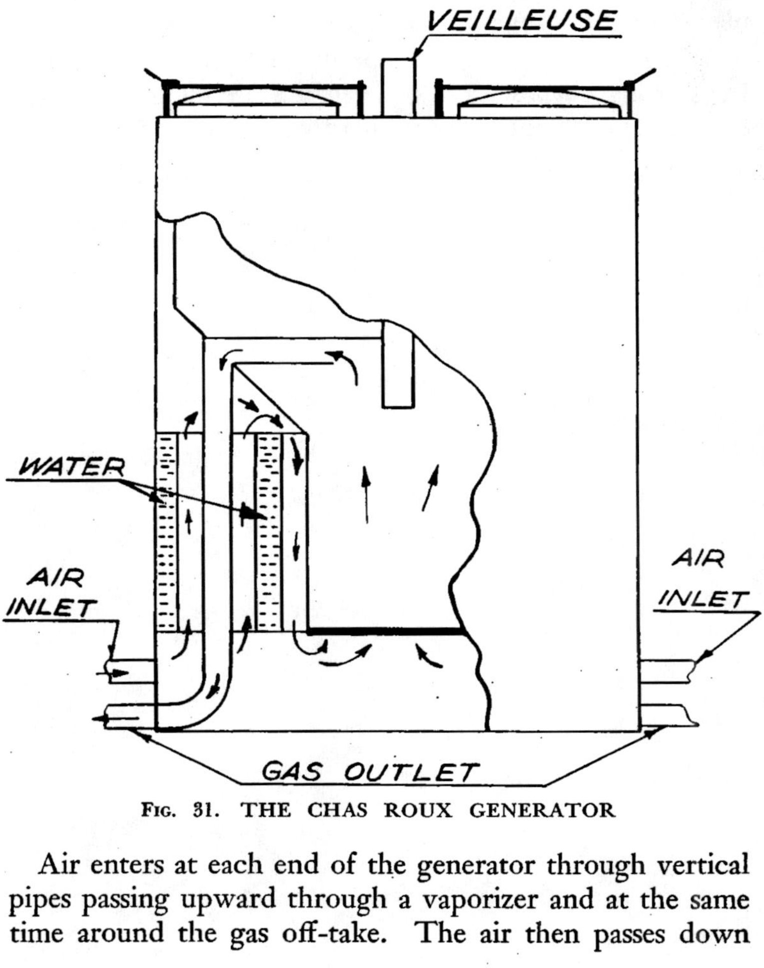

FIG. 31. THE CHAS ROUX GENERATOR

Air enters at each end of the generator through vertical pipes passing upward through a vaporizer and at the same time around the gas off-take. The air then passes down through a narrow space between the vaporizer and the furnace walls into the ash box and up through the grate on which the fuel is burning. Gas leaves the top of the fuel bed and passes through two pipes—each surrounded by an annulus, through which the ingoing air is passing in the opposite direction—and is then connected in one main, leading under the vehicle to the cleaner, which is placed on the opposite running-board.

There is a small central pipe or veilleuse which is used for starting purposes and to keep the fire alight when the engine is not running. This is, of course, closed when the generator is in use.

CHAPTER XIII

ENGLISH PRODUCER GAS PLANTS

The Brush-Koela Plant

At the time of writing the Brush-Koela gas producer is one of the plants being made on large scale production in England, and for this reason, as well as for its unique design, it will be considered here. Koela is an Indian word for charcoal, it being in India that the plant was first developed. At the outbreak of the present war the manufacturing rights were obtained by the Brush Electrical Engineering Company of England.

The plant consists of a generator and two scrubbers connected as shown in Fig. 32. The generator is of novel design, in as much as it incorporates both the cross-draught and up-draught principles, the former being used for starting and the latter system brought into operation for regular use. The change from one system to the other is made by a change over valve. Steam is admitted to the fire by being first vaporized by the hot exit gases. The fire grate is of the hanging basket type, and the starting tuyere is composed of carborundum in order to avoid the necessity of providing a water-cooling system which would only receive occasional use.

The primary scrubber has a series of apertured baffles and both a fibre and wood-wool filter element. The secondary scrubber is composed of still more baffles and filters, as shown in Fig. 32. Although this plant is operating on anthracite, it would be still more effective on the highly reactive charcoal for which Australian plants are designed.

It is of interest to note that the price of the plant, excluding the installation costs, is £95 (English).

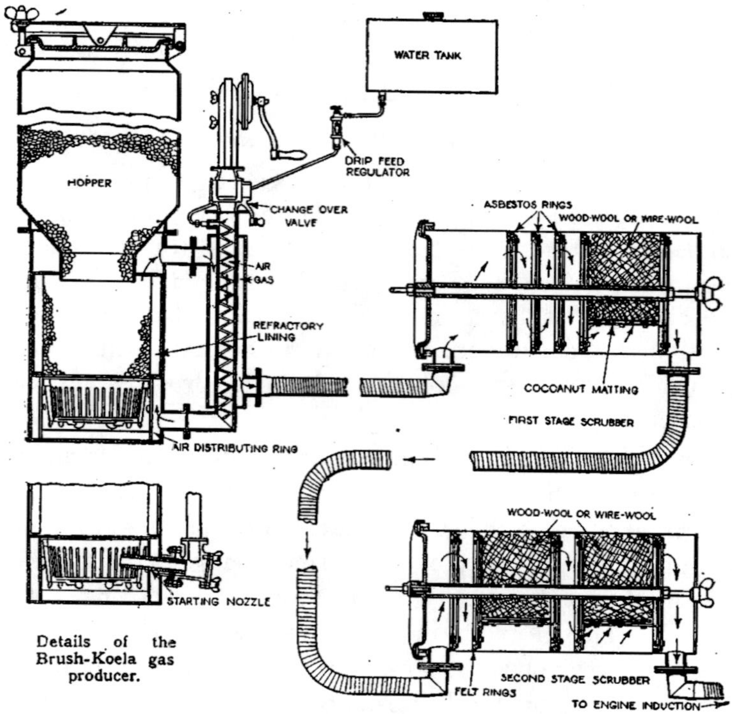

FIG. 32. THE BRUSH-KOELA GAS PRODUCER

THE NEIL AND SPENCER PLANT

The Neil and Spencer generator is of the orthodox cross-draught type and uses a tuyere which is water-cooled from an additional radiator mounted near the plant. This is found necessary owing to the higher temperatures obtained within the generator when anthracite (the fuel for which most English plants are designed) is used.

The cleaner consists of a layer of small broken coke soaked in oil, to extract the dust, and a layer of cotton waste, to remove moisture. Before passing to the scrubber the gas passes through a long horizontal cylinder, the first portion of which contains a centrifugal cleaner; the other

FIG. 33. THE NEIL AND SPENCER GAS PRODUCER

1. The generator. 2. Air inlet to tuyere. 3. Centrifugal cleaner and expansion cooling chamber. 4. The final scrubber consisting of oil-soaked coke and cotton waste.

section is an expansion chamber into which the gases are expanded, this having the effect of lowering their temperature.

Fig. 33 shows a Neil and Spencer plant mounted on the back of a car. Such a plant as shown weighs 2½ cwt. when fully loaded with fuel.

CHAPTER XIV

AUSTRALIAN PRODUCER GAS PLANTS

The Powell Producer Gas Plant

The Powell gas plant consists of a generator of a modified up-draught type, two cyclone primary cleaners, and an oil scrubber with fibre filter.

The generator consists of a double-walled steel cylinder (see Fig. 34). The annular space between these two walls is divided approximately at centre by a steel rim which is welded to the lower or firebox section.

Air for combustion enters through holes (A) in the outer sleeve just below the dividing rim, and is drawn down through the annular space between the firebox and the outer sleeve, thus cooling the walls and at the same time preheating the air. Water is injected into this space, where it is vaporized and passes with the incoming air through the grate to the fire.

The grate, which is a dome-shaped casting, has four orifices near top centre through which the air enters and converges in the centre of the firebox to produce a centralized fire of great intensity. Owing to the shape of the grate, any clinker formed flows by gravity to the lower level where it does not interfere with the fire or clog the grate.

The gas (a combination of air and producer gas enriched with water gas) passes up to the slit (E) in the inner sleeve, where it passes out into the annular space between the two sleeves. Here it is partially cooled by radiation from the inner sleeve to the cooler charcoal in the top of the generator and from the outside surface to the air. It

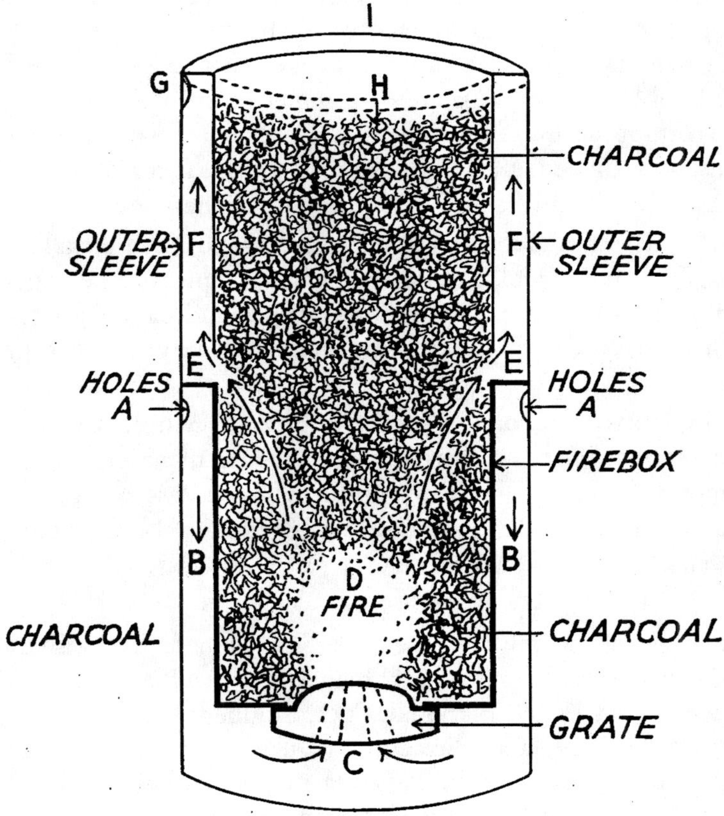

Fig. 34. SKETCH SHOWING THE PRINCIPLE OF THE POWELL GENERATOR

is then drawn off through three evenly spaced holes in the top of the outer sleeve to the primary cleaner.

This cleaner (and cooler) is of the cyclone type, in which fixed vanes impart a rotary motion to the gas, throwing the dust particles to the outer edge while the gas is taken off at centre. The second dry cleaner is of a type patented by Dr Bowden of the University of Western Australia, of which Powell Gas Producers Ltd has the patent rights. It is claimed to be the most efficient dry cleaner on the market. In the oil cleaner the gas is caused to bubble through a gauge pad immersed in oil, producing a frothing action, in which it is claimed the last trace of dust is removed. The gas is freed from traces of oil by means of a pad of sisal before passing to the engine.

A large number of Powell plants have been fitted to tractors and trucks. More than 500 plants are now (1 April 1940) fitted to tractors, trucks and cars throughout Australia, with (the makers claim) entirely satisfactory results.

In January 1940, 200 of the 400 plants whose sale was guaranteed by the Commonwealth Department of Supply were allotted to this firm whose main address is 142-146 Parramatta Road, Camperdown, Sydney, New South Wales.

The Wishart Gas Producer

A very modern and advanced producer is made by Wishart Gas Producers Limited of Blackburn, Melbourne, and, under licence, by National Fuel Engineers of 3 Spring Street, Sydney, and Buzacotts (Queensland) Limited of Brisbane. It is made in standard models of two sizes for motor trucks, and of three sizes for motor cars.

The producer is of the cross-draft type with water-cooled tuyere and a patented steam injection system which uses the steam generated in the tuyere and requires no adjustment while in operation. The section illustrated in Fig. 35 shows the truck producer, the lower portion of which forms a box frame of great strength. The essential parts—charcoal hopper, generator, steam system, water tank, dust trap, cooler, and scrubber, all of generous size—are arranged into one light-weight, compact unit to fit quickly to the front of the motor truck. In this position, it does not interfere with the cabin or loading space, and any heat generated does not come into contact with the load carried.

This producer is entirely self-regulating, and has an exceptionally long operating range. An automatic air and gas mixing valve (described elsewhere) keeps the mixture in correct proportions under all conditions, and a special idling screw is fitted. These features make the producer particularly easy to handle in traffic.

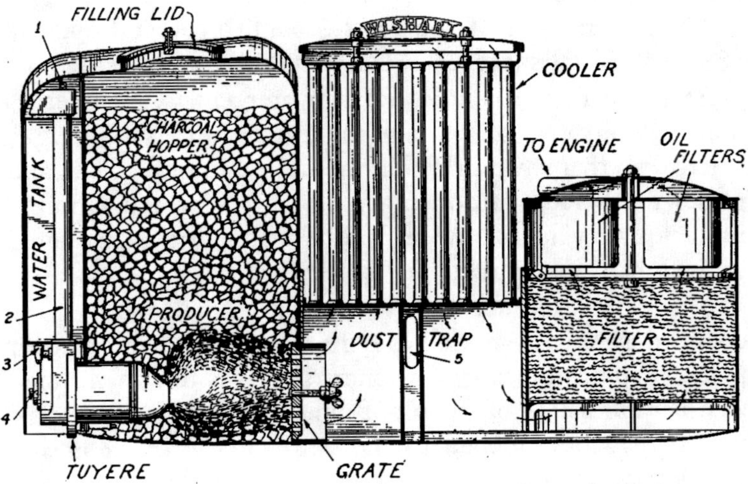

FIG. 35. WISHART PRODUCER GAS PLANT, TRUCK TYPE

Special attention is paid to the cleaning of gas, which passes first through sisal hemp and then through a special oil bath type of filter to remove the last traces of dust from the gas.

The car producer is similar to the truck producer

described above except for a rearrangement of parts to enable the plant to be neatly fixed to the rear of the car.

A six-volt electric blower is available for starting up without the use of petrol; or starting may be effected by the usual method of running the engine on petrol.

The Wishart producer is also made in models to suit farm tractors, motor boats, portable engines, and stationary engines, in sizes from 1½ h.p. upwards.

The Fleetway Gas Producer

The Fleetway gas producer is made by Messrs Fleet Forge Pty Ltd of 9 Lorimer Street, South Melbourne. It comprises a generator, primary cleaner, cooling radiator, secondary cleaner (Wright cleaner) and a mixing chamber.

The generator is of the cross-draught type and consists of a rectangular sheet metal container 18 x 18 x 48 inches.

Attached to the generator is a water tank which serves the dual purpose of cooling the tuyere and supplying a water drip that is vaporized in the heated tuyere orifice and passes to the fire, producing a water gas which enriches the calorific value of the gas. The water drip is regulated by a sight drip regulator in the driver's cabin.

A primary cleaner of the cyclone type removes the coarse dust and is emptied as required by removing a cover plate. From the primary cleaner the gas is drawn through a cooler consisting of a bank of pipes connected by headers, which may be attached to the body of the truck in any location preferred.

After cooling the gas passes to the secondary cleaner, the Wright dust extractor, which consists essentially of a water impingement cleaner, a cyclone and sisal cleaner, housed in the one composite unit.

After leaving the Wright cleaner the purified gas is taken to a mixing chamber attached to the engine manifold at the carburettor, and to which the carburettor is connected by means of a branch pipe. Here the correct air-gas mixture is automatically regulated and passes through the accelerator-controlled butterfly valve to the engine cylinders.

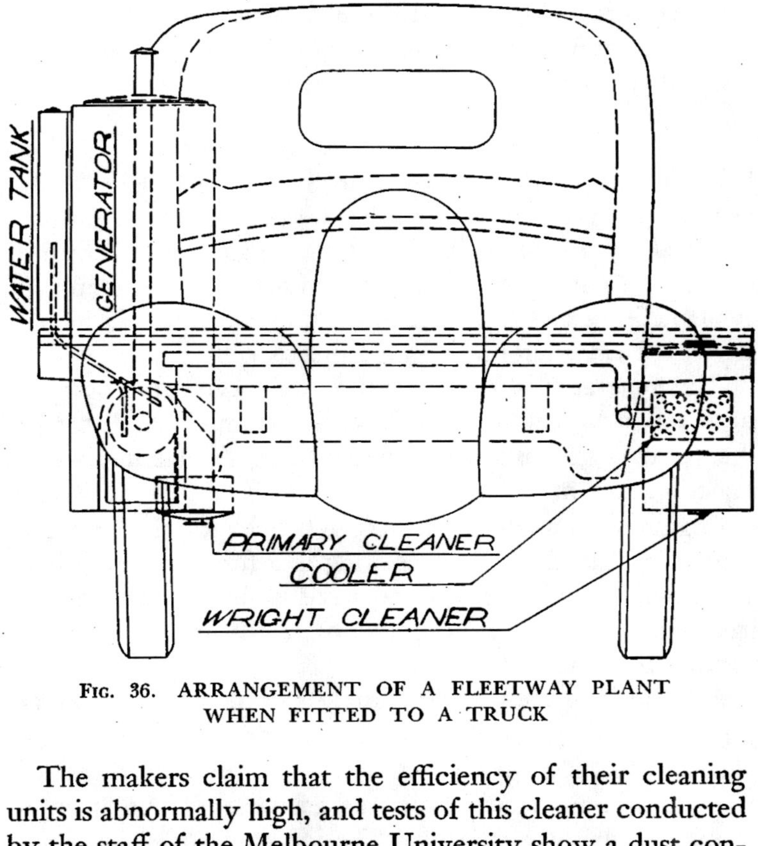

FIG. 36. ARRANGEMENT OF A FLEETWAY PLANT WHEN FITTED TO A TRUCK

The makers claim that the efficiency of their cleaning units is abnormally high, and tests of this cleaner conducted by the staff of the Melbourne University show a dust concentration of only 16 to 18 milligrammes per cubic metre of purified gas. (The specification of the Department of Supply and Development stipulates a maximum dust concentration of 200 milligrammes per cubic metre.)

Cleaning is effected without any noticeable back pressure and power curves of engine performance on producer gas with Fleetway plants are amongst the best recorded for the compression ratio employed.

Plants are available for trucks, tractors and cars and carry the guarantee arranged by the Commonwealth Department of Supply and Development.

Fig. 36 shows the arrangement of a Fleetway plant when fitted to a truck.

The Panther Producer Gas Plant

Panther gas producers, a number of which have been ordered by the Commonwealth Government, are manufactured by Messrs Carbo-Gen Gas Producer Company Pty Ltd, 44 Margaret Street, Sydney. The generator is available in three sizes: 3 feet, 4 feet, or 5 feet, by 22 inches square, constructed of 12-gauge sheet steel.

The generator is of the cross-draught type employing a tuyere which is water-cooled, either from the radiator of the vehicle or from a special water tank. A patented steam injection system is employed.

Engine suction draws air at high velocity through the tuyere into the producer, resulting in a fire zone of limited area which maintains a constantly high temperature irrespective of the throttle opening, a feature essential for flexible performance.

After leaving the generator the gas passes to the scrubber through a long length of steel tubing which takes the place of a special cooling unit. The cleaner is of the liquid impingement type, the gas passing first through water and then through oil. The incoming gas is directed by a baffle on to the water after which it passes through a layer of moist cleaning material, and then to the final oil bath.

The Panther automatic air and gas mixing valve is shown in Fig. 37. This valve is constructed so as to provide the correct mixture of air and gas for the engine, whether idling, accelerating or maintaining a constant speed. As seen in the picture, the mixing valve consists of a manifold

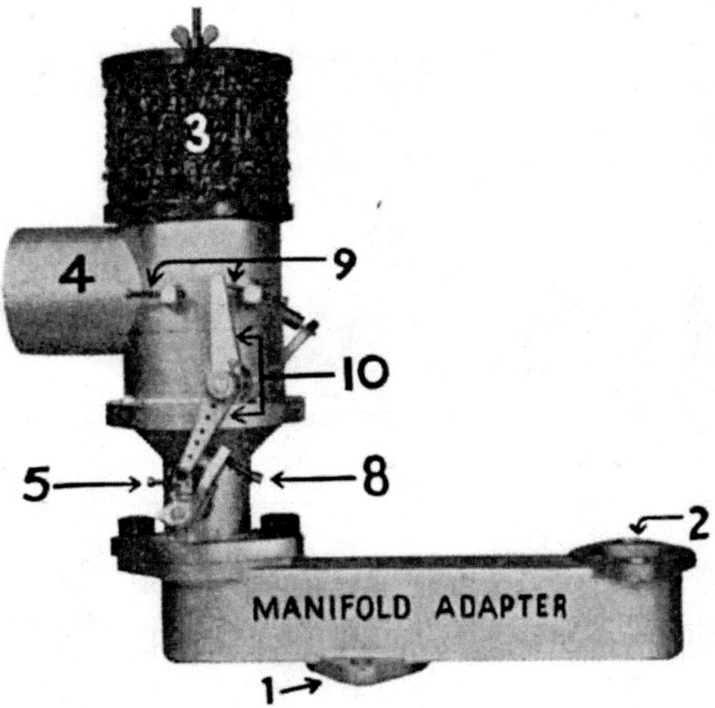

FIG. 37. PANTHER MIXING VALVE

adapter which is mounted on the engine manifold at 1 and has the petrol carburettor attached at 2, while to the other end is bolted the actual mixing valve. This valve has an idling speed adjusting screw (5), an idling mixture adjusting screw (8), and a high speed mixture adjusting screw (9).

The correct air-gas setting is obtained by adjustment of the linkage arms (10). In operation the gas enters at 4, and the air through an oil-soaked copper cleaner at 3. The two then mix and pass to the engine in the desired quantities. If it is desired to use petrol, the gas butterfly valve is closed and the carburettor used in the ordinary manner.

Panther gas producer plants are available for use on trucks, tractors and private cars. In the latter case they are mounted on a single castor wheel trailer.

Fig. 38 shows a Panther plant fitted to a truck.

FIG. 38. TRUCK FITTED WITH PANTHER GAS PRODUCER PLANT

HIGH SPEED GAS

The well-known English High Speed Gas plant is being manufactured in Australia under licence by Messrs Malcolm Moore Ltd, of Williamstown Road, Port Melbourne, who supply the following particulars (see Figs 39 and 40):

This plant consists of three main components: (i) the producer, (ii) a cooler and scrubber, and (iii) a gas mixing valve.

The producer is a suitably shaped mild steel generator which has no refractory lining, and in which the path of gas is in a horizontal direction (cross-draught). Air to water is drawn through a water-cooled tuyere, and the gas

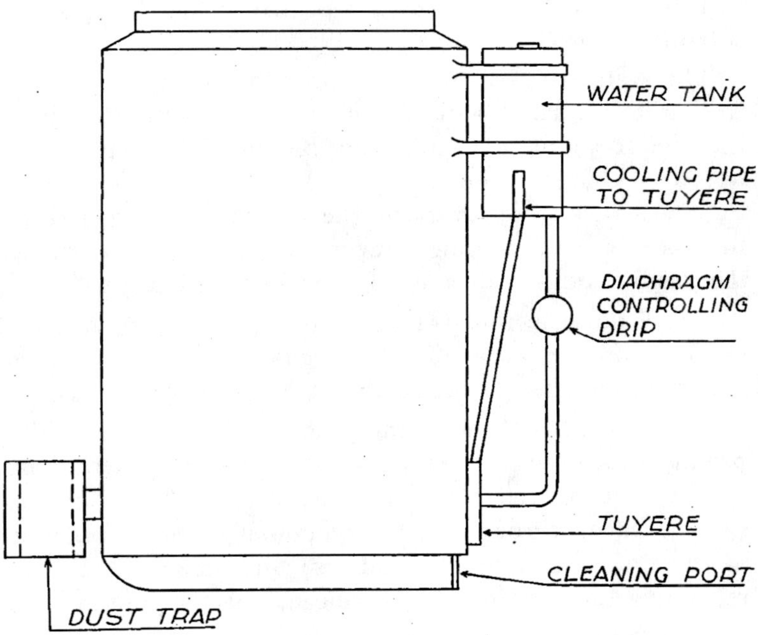

Fig. 39. THE HIGH SPEED GAS GENERATOR

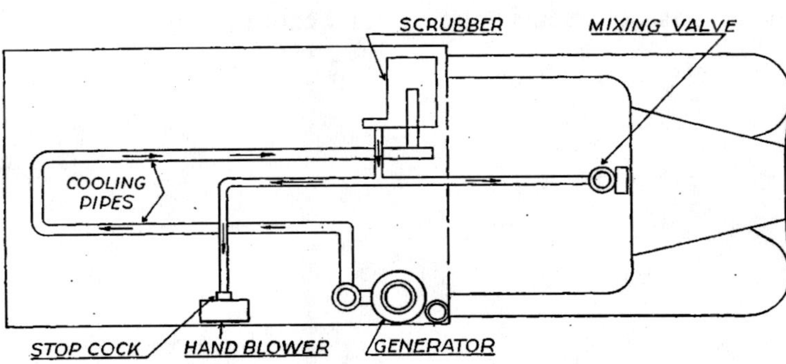

Fig. 40. LAYOUT OF THE HIGH SPEED GAS PLANT

K

generated in the furnace zone is drawn off on the opposite side of the producer.

The water that is drawn in through the tuyere is controlled by a small valve which will regulate the drop required to about 10-15 drops per minute. This water drop does two things in the fire zone: (i) It surrounds the fire by a wet belt which restricts the fire to such a degree as to make refractory lining unnecessary; and (ii) it enriches the gas by addition of a small quantity of hydrogen.

It will be seen, therefore, that the highest temperatures are right in the main air stream and liberation of gas is almost instantaneous—hence the term "high speed gas."

The gas passes through the grate set in the side of the producer into the primary expansion chamber, where most of the large particles of carbon are deposited. From this the gas travels through a nest of cooling pipes, and thence to the final sisal scrubbers. These units take out all the remaining dust and deliver the clean, low temperature gas to the engine.

The gas is mixed with its correct amount of air at a mixing valve mounted alongside the standard carburettor, and is controlled by the standard foot accelerator mounted in its usual location in the driver's cabin.

CHAPTER XV

BUILDING YOUR OWN PRODUCER GAS PLANT

WHILE people in general would be well advised to purchase their gas plants from a well-known and reliable manufacturer, rather than make their own, it is certain that a number of enthusiastic amateurs as well as experienced garagemen will make the attempt. To these the present chapter will be of interest.

Many farmers' sons, accustomed to the management and repair of tractors, trucks, and farm machinery, are no mean mechanics, and the farm workshop is often quite well equipped. Oxy-welding equipment is, however, not generally included in the equipment, and without a welding plant the manufacture of a really serviceable gas producer is almost impossible. The following instructions, therefore, assume that the services of a welder are available. In the home workshop, where this is not the case, welding can be reduced to a minimum, and the vital parts taken to the nearest garage or welder.

The plant described has an up-draught generator, of which a considerable number have been used for some years and have stood the test of actual service. This type, while lacking the flexibility of a cross-draught generator, presents fewer difficulties for the home manufacturer who could not afford the cost of patterns for tuyere etc. Moreover, the application of steam is simplified and the performance of the plant when properly made is excellent.

It is assumed that the engine is of approximately 25 h.p.

and has a compression ratio of not less than 4½:1. As stated elsewhere, a higher compression ratio (up to 8:1) is desirable.

The generator is, as we know, made up of several parts and is shown in Fig. 41. The body is a cylindrical container 3 feet 6 inches high and 15 inches in diameter, made

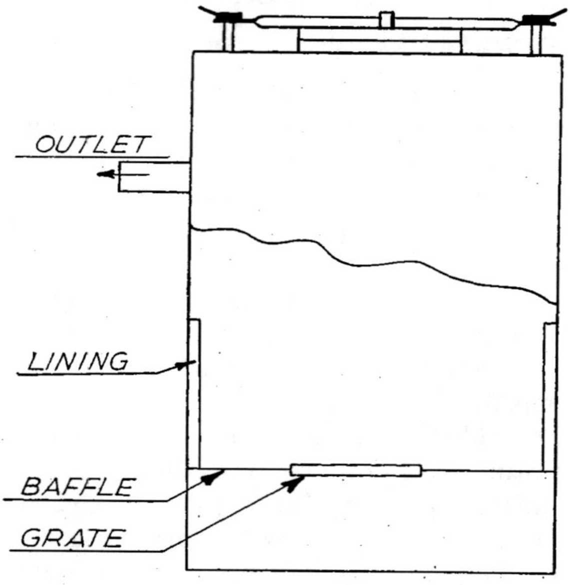

FIG. 41. THE GENERATOR

of not less than 20-gauge mild steel. The home manufacturer will find that two 12-gallon oil drums with the tops removed, which can be welded together at the centre line after fitting the various accessories, will serve excellently.

Each part will now be considered separately.

BODY OF THE GENERATOR

As previously stated, the body consists of two 12-gallon

oil drums from which the tops have been removed. Before welding these together the various parts will have to be fitted. A circular hole 12 inches in diameter is made in the bottom of the drum which now becomes the top of the generator, and a circular rim of $\frac{1}{8} \times 1$ inch hoop steel is bronze-welded to it to provide a seating for the lid.

To make a door for the removal of ash, a cut-out of 5×3 inches is made 4 inches from the bottom of the lower drum. A steel rim of $1\frac{1}{2} \times \frac{1}{8}$ inch is bronze-welded around this cut-out.

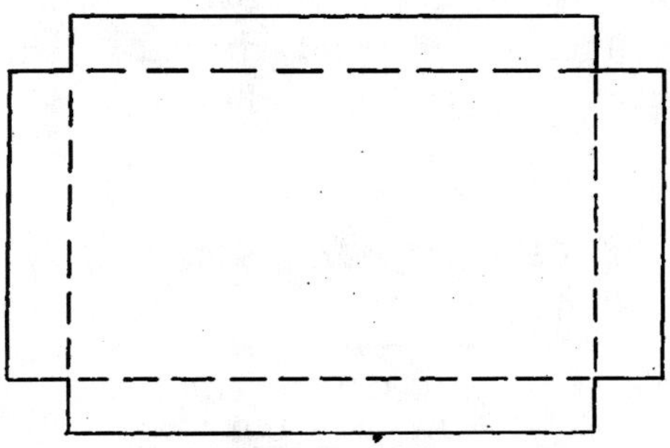

FIG. 42. BLANK OF ASH AND FIRE DOORS

Ash Door

The ash door (Fig. 42) consists of two plates, 6×4 inches for the outer and 5×3 inches for the inner, each turned down $\frac{3}{8}$ inch around the edges to provide a space to carry a $\frac{1}{2}$-inch asbestos rope. This door will have to be made from 18-gauge mild steel.

Fire Door

A fire door of the same construction as the ash door may be situated 1 inch above the grate baffle. Although this is optional, it is very desirable for maintenance and cleaning; but it is important that it should be air-tight. The omission of this door is advised if the necessary equipment is lacking.

Holding-down Bar for Fire and Ash Doors.

This bar is made from a piece of mild steel, $1\frac{1}{2} \times \frac{3}{8}$ inch, and is 10 inches long. It has a $1\frac{1}{2}$-inch slot in each end and is welded to the door. Two holes are placed in the generator to take $6 \times \frac{1}{2}$ inch bolts, which engage in the slots in this bar and so hold it down. Wing nuts are used for tightening.

The Lid

The lid for the generator is made from two plates of $13\frac{1}{4}$ and $12\frac{1}{4}$ inches in diameter respectively, turned down for $\frac{3}{8}$ inch around their outer edges, riveted together and (as Fig. 43 shows) carrying an asbestos rope of $\frac{1}{2}$-inch

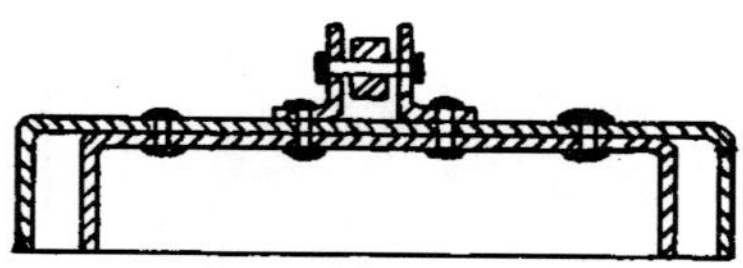

FIG. 43. THE GENERATOR AND SCRUBBER LID

diameter. Two $\frac{1}{8}$-inch diameter holes are bored 1 inch each side of the centre line to take the strips for the holding-down bar. These strips, two of which are required, are shown in Fig. 44. They are made from $\frac{1}{8} \times 1$ inch copper, and have holes bored in them where shown.

Holding-down Bar for Lid.

The holding-down bar for the lid is of $1\frac{1}{2} \times \frac{1}{2}$ inch mild steel; it has a $\frac{5}{16}$ -inch hole in the middle and a $1 \times \frac{1}{2}$ inch slot in each end, which is then twisted at 90 degrees.

Two $\frac{1}{2}$-inch diameter holes are bored in the top of the drum to take $6 \times \frac{1}{2}$ inch bolts, which engage in the slots of the holding-down bar and enable the lid to be tightened down. It will be found advisable to have a $\frac{1}{4} \times 4$ inch piece of mild steel rod welded to each of the nuts to provide easier tightening. The lid will have to be made from 12-gauge steel.

Grate Baffle

The circular grate baffle is provided to carry the grate and act as a bed for the fuel. The baffle should be made from a piece of 16-gauge steel of 16½ inches diameter, having ¾ inch turned down around the edge. An 8-inch centre is

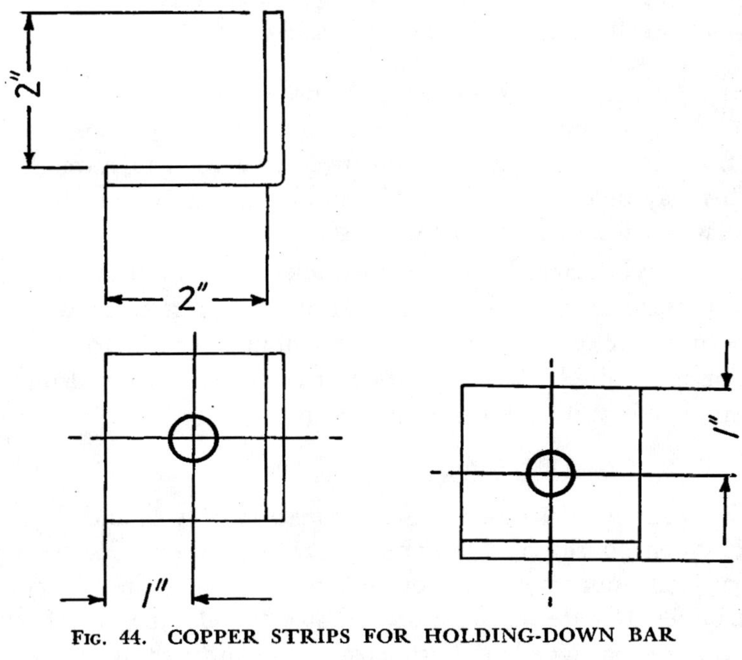

FIG. 44. COPPER STRIPS FOR HOLDING-DOWN BAR

removed to carry the grate. The baffle is to be riveted to the side of the bottom drum at a distance of 9 inches from the bottom.

The grate should be 8 inches in diameter with ⅝-inch opening and ⅝-inch bars. A casting of this will have to be obtained from an iron foundry.

Air Intake Pipe

An air intake pipe of 1½-inch steel tubing approximately 1 foot long is fitted into a right angle bend, which is in turn inserted into the drum 1 inch below the grate baffle.

Gas Take-off

A long piping bend 1½ inches in diameter with a 2 × 1½ inch reducing socket is screwed in the centre drain plug of the top drum and provides the gas take-off. To prevent the fuel being carried away in the gas stream two bars of ⅜-inch steel are welded across the socket.

Refractory Lining

In order to prevent burning of walls of the generator a refractory lining will be found necessary for the furnace. This may be obtained from Messrs Non-Porite Pty Ltd, of Burwood Road, Hawthorn, Victoria.

This cylindrical liner, 1 inch thick, is made up in three sections tongued and grooved for fitting together, with aperture for grate, and with or without aperture for fire door as required. This should be fitted to the bottom drum before it is welded to the top section.

The Scrubber

The scrubber described is a combination of wet and dry, of which considerable numbers have been used. Its construction presents a few more difficulties than a simple dry scrubber, but the use of water allows the elimination of a cooling unit, which is both costly and difficult to make. The greater majority of the work is soldered, only a few welded parts being used.

The body, lid and baffles of the scrubber are made from a sheet of 20-gauge galvanized iron 6 feet by 2 feet 6 inches, which can be purchased fairly cheaply.

Since the unit is round it may be found advisable to have it rolled by someone possessing the necessary equipment, although this could be done at home if sufficient care and time were taken. Each part is treated separately. Take your flat sheet and mark out on it a rectangle 30 × 38½ inches, so that when this is rolled and the edges turned down to form a lock joint a cylindrical container 12 inches in diameter and 29¼ inches high will result.

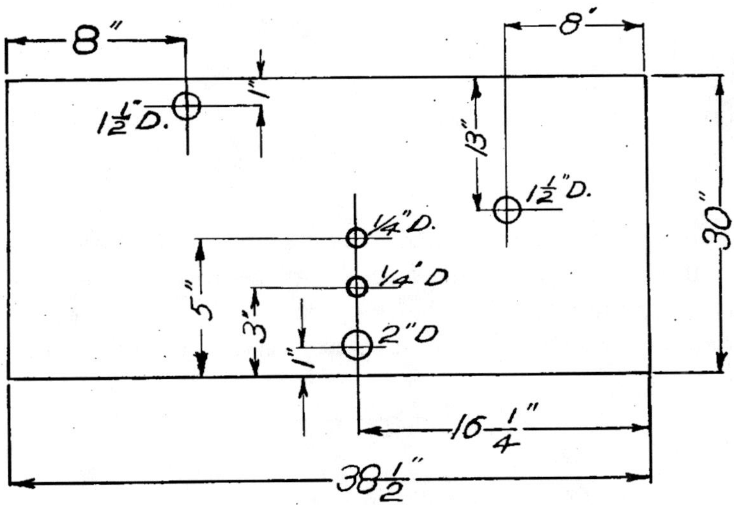

FIG. 45. SCRUBBER BLANK

Fig. 45 shows the position of five holes which have to be cut out. It is much easier to cut these holes before rolling the metal. When the holes are made the flat sheet of metal is rolled into the cylindrical shape.

Down the sides there will be two edges of metal meeting and, in order to provide a stronger container and to make the soldering easier, these edges should be lock-seamed. The edges are turned down in the opposite direction for about $\frac{3}{8}$ inch so that they slip into one another; and when they are tightened down and then soldered a very good joint is made. Fig. 46 shows the principle of the lock joint. Solder the side seams to about 1 inch from each end and then roll over a $\frac{1}{4}$-inch flange at the top and bottom. Fig. 47 shows the principle of the method of joining.

FIG. 46. THE LOCK SEAM

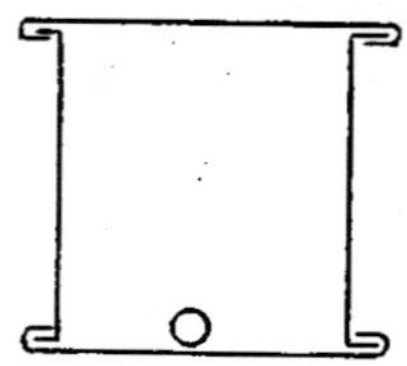

FIG. 47. METHOD OF JOINING SCRUBBER PARTS

This flange increases the diameter of the cylinder to about 12½ inches. The pieces for the top and bottom (Fig. 48) must therefore be 13¼ inches in diameter in order to allow a $\frac{3}{8}$-inch lip to be turned over to lock these portions into position; they must then be soldered down to prevent leaks, and the side seams finished.

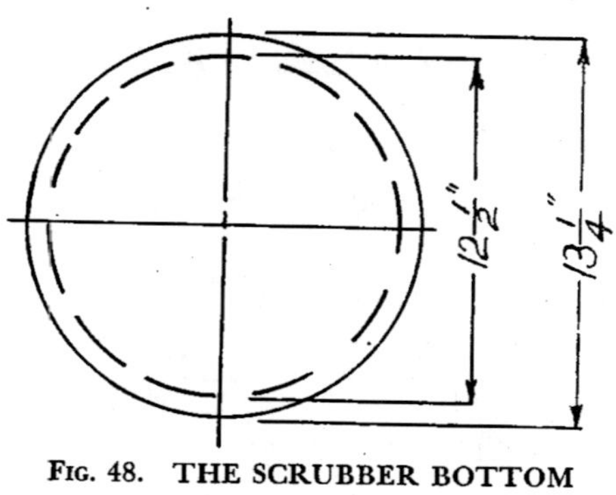

FIG. 48. THE SCRUBBER BOTTOM

Before fitting the top cover, the baffle must be soldered

into position 12 inches below the top edge. This baffle has a maximum number of ¾-inch holes in it, and its purpose is to act as a support for ¼-inch mesh gauze which fits over it and so prevents the sisal falling through into the water container.

The top cover for the scrubber, into which the removable lid fits, is shown in Fig. 49. A hole 8 inches in diameter is cut out of this and, as was the case with the generator, a 1 × ⅛ inch loop steel rim is welded into it. Two 4 × ½ inch brass bolts are soldered into the scrubber top to engage in the slots of the holding-down bar.

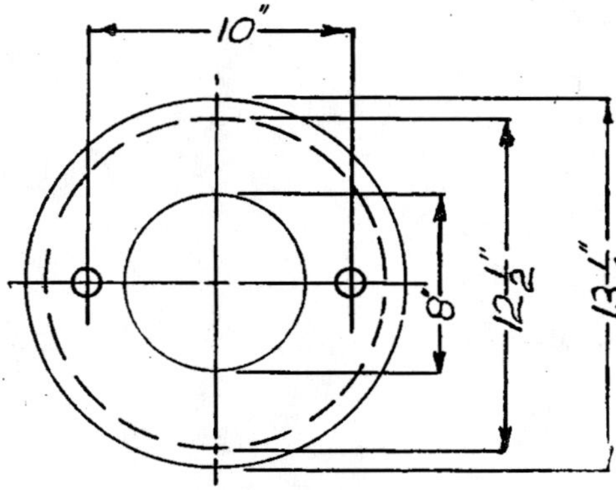

FIG. 49. THE SCRUBBER TOP

A lid identical in shape with that of the generator is used. It is, however, made of two plates of 9¼ inches and 8¼ inches in diameter, which are turned so as to leave the usual ½-inch space between them, into which the ½-inch diameter asbestos rope fits.

Gas Spreader for Scrubber

The gas spreader consists of a cylindrical blank 10 inches in diameter of 18-gauge mild steel, serrated as shown in

Fig. 50. The serrations, which are ½ inch deep, are turned down to provide saw teeth, by which the gas is split up into small streams, so making for better cleaning.

In the centre of the blank is cut a slightly elliptical hole through which the lead-in pipe projects approximately ¼ inch, and to which the spreader is bronze welded. This baffle should be placed about on the same level as the lower water level indicator.

The lead-in pipe needs to be about 24 inches long and to the end of it is welded a steel flange as shown in Fig. 52.

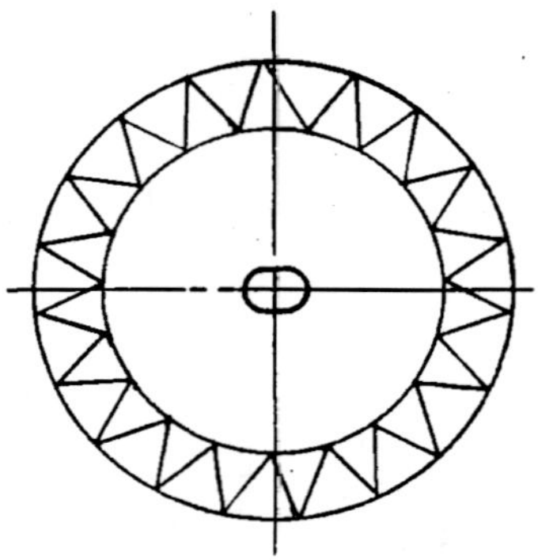

Fig. 50. THE GAS SPREADER

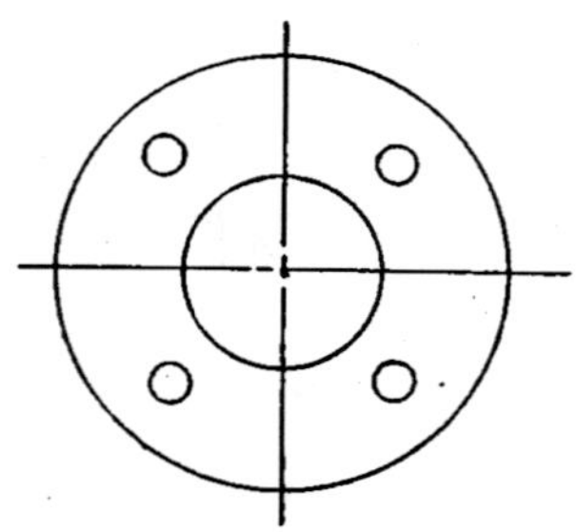

Fig. 51. FLANGE FOR LEAD-IN PIPE

This flange is a 1½-inch I.D. washer, having four holes drilled in it (Fig. 51) to match the flange from the generator off-take. The intake pipe may be supported by means of a band around the scrubber.

A drain plug of 2-inch diameter is provided and a brass fitting screwed with 2-inch piping thread is soldered into the hole provided. Two fittings for ¼-inch piping plugs are soldered into the holes provided to act as water level indicators. The location of these three holes in the blank is shown in Fig. 45.

The gas off-take from the scrubber is provided by 18-gauge seamless steel tubing, 9 × 1½ inches, which is bronze-welded to the scrubber 1 inch from the top, as indicated in Fig. 45. The inner end of this tube is provided with a ¼-inch mesh gauze cover to prevent the sisal being carried down the line.

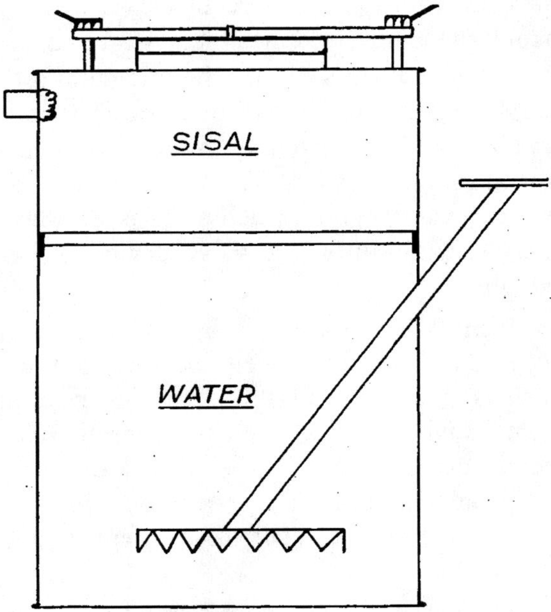

FIG. 52. THE SCRUBBER

The gas is carried to the engine manifold by means of a length of 1½-inch rubber suction hose which is obtainable from any rubber company.

FITTING TO ENGINE

Having decided on the location of the various units, next choose the point of entry of the gas into the inlet manifold or induction pipe. At the chosen point a branched pipe is brazed in, one branch carrying gas from the generator and the other the air, as shown in Fig. 18. To the air branch is fitted the automatic air regulator which may consist of a spring regulated valve of very light construction, usually a light mushroom valve sitting on a flat seat. The air entry is of about 1½ inches diameter and the mushroom valve a little larger. In addition, an extra air entry, manually controlled from the dash by a flexible wire control, can be fitted; it is usual to fit this so that the minimum setting just gives enough air for idling with the mushroom valve closed. The extra air should be capable of giving more than enough air even at maximum power. This extra air device may be of the slotted band type, which clamps around the air branch between the automatic valve device and the joint to the gas pipe.

The butterfly valve shown in Fig. 18 thus controls the amount of mixture entering the engine, but not the strength of the air-gas mixture. This butterfly is best provided with a foot pedal control, which is better if independent of the petrol controls; it may be placed to the right of the brake pedal and near the petrol foot pedal, but not so close that the two are likely to be depressed together inadvertently.

To determine the right size of gas-air butterfly valve, measure the petrol-air butterfly in the carburettor and make the gas-air butterfly ⅛ inch bigger in diameter; the pipe must accommodate this size. It is worth while to fit a 2-inch water tap of the "full way" type between the manifold and the gas-air butterfly, although this involves extra trouble and expense. This tap, which must be of brass, remains always wide open except when it is desired to change back temporarily to petrol; then it is turned off, so that no air can leak into the manifold from the idle gas system.

If desired the automatic air mixer and extra air device can be fitted on the dash, and the air from the mixer carried by a 1-inch rubber hose to the air branch pipe near the manifold. On bigger engines 1½-inch rubber hose may be found necessary.

CHAPTER XVI

OPERATING COSTS OF PRODUCER GAS

THERE are few things upon which there is so much room for diversity of opinion as the operating costs of a gas producer. The large number of varying factors that come into operation make it virtually impossible to predict these costs with any accuracy. Among these factors are the initial cost of the plant, the cost of fuel, the make of plant, the make of vehicle, the type of load and kind of work the vehicle will be performing, and the kind of country it will be operating in.

All we can hope to do here is to consider the operating costs of a number of vehicles that have been tested from this viewpoint. From these figures it should be possible for the individual to make a reasonable assumption to suit his own particular case. It must be remembered that the costs set out below are those obtained on long distance runs, in which economy is considerably greater.

It is a rough and ready rule among gas producer operators that a corn sack of charcoal (hard wood) will provide the same power output as four gallons of petrol. In actual practice one has to make allowance for the fact that with frequent stopping and starting a proportion of the charcoal is wasted in heating up the generator.

When all of these factors are considered, charcoal is undeniably the cheapest fuel available for motor transport.

In determining running costs for producer gas, there are a number of factors often overlooked. Among these is the additional time required for producer gas—time which would have to be paid for if the vehicle were being driven by a hired hand. It is common practice for tests to be made between two towns, a log being taken of the distance travelled, the speed obtained, the amount of charcoal used and the amount of petrol used. The cost per mile is then computed and presents a very low and attractive figure. However, it is often overlooked that at the end of the journey (if it is a long one) the scrubber has to be cleaned and the generator emptied and refilled for the next run. Over an average period of several months, an English firm of fleet operators found that the extra maintenance time required for their Gohin-Poulenc plant was a half-hour per day.

Over and above the maintenance required on the gas producer, an engine running on gas requires to be overhauled nearly twice as often as with petrol, and even with the best producers the life of the engine will be lessened, although by not nearly so much as is popularly imagined.

At the time of writing the fitting of a gas producer results in an increase in the registration cost of a vehicle, owing to the added weight. It is to be hoped that this state of affairs will not continue to exist, but rather that a bonus be given to encourage the use of producer gas.

The cost of fitting a plant will be determined to a large extent by how long the plant lasts, which in turn depends on the type of plant.

A gas producer usually fails owing to burning out of the generator, rusting of the scrubber unit, or breaking of joints and seams by vibration. Of these troubles, burning out of the generator and tuyere is the worst; it is often due to an attempt to reduce the weight of the plant by using too thin a metal in their construction, and using too little water during their operation. If the plant is insecurely mounted and the piping not rigidly supported, leaks may be set up which will cause a great deal of trouble during running and will eventually demand replacement of the piping and repairing of the seams and joints.

Some manufacturers are claiming that their plants will last as long as the car to which they are attached. We consider this to be an optimistic claim.

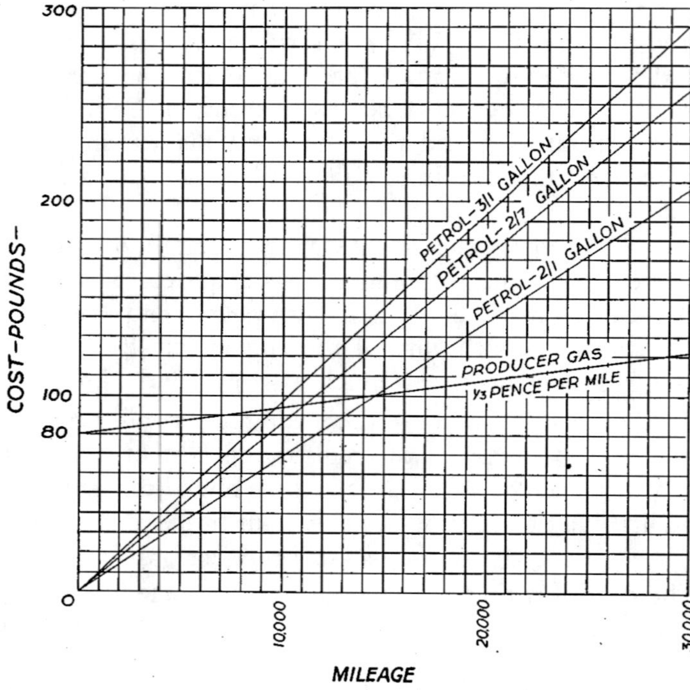

Fig. 53. GRAPH SHOWING RELATIVE COSTS AND MILEAGE OF PETROL AND PRODUCER GAS

The graph (Fig. 53) is designed to show just what mileage must be travelled on producer gas in order to save the initial cost of the producer gas unit, and to show the savings effected thereafter.

Where the producer gas line crosses each petrol line indicates the mileage necessary before the purchaser has recovered his outlay on the plant at the different petrol costs. The purchaser has at that mileage only spent as much in all as if he had not changed to gas, and has a good second-hand plant paid for out of his savings.

The producer gas running cost of ⅓ of a penny per mile includes extra cost for servicing and maintenance. It is assumed that the same vehicle would run fifteen miles to the gallon on petrol. It is also assumed that the cost of the plant was £80 and that charcoal cost £3 10s. per ton.

The following results were obtained in a test run of a gas producer truck from Canberra to Melbourne and back. The test was carried out by the Department of the Interior and the vehicle had been in use for seventeen months. On the journey from Canberra to Melbourne the truck carried a nett load of over 6 tons, and 2½ tons on the return trip. For the whole journey of 975 miles, which was run at an average speed of 27 miles per hour, the fuel consumed was 1254 lb. of charcoal at a cost of 0·573 pence per mile. One quart of lubricating oil was also used.

Table VIII
OPERATING COSTS OF PRODUCER GAS UNDER VARIOUS CONDITIONS

Authority	Time Taken for Journey	Distance Travelled (miles)	Make of Vehicle	Charcoal Used	Load Carried	Petrol Used	Cost of Charcoal per 50 lb. Bag	Cost per Mile on Producer Gas (pence)	Cost per Mile on Petrol (pence)	Average Speed (m.p.h.)
Wishart ..	14 hrs.	571	1934 8-cylinder Pontiac coupe.	450 lb.	—	—	1s. 6d.	·284	—	41
Mr J. Fielder (N.R.M.A.)	—	55·5	1937 Oldsmobile car.	47 lb.	—	—	1s. 4d.	—	1·39	40
Carbo-Gen	160 min.	42	2-ton Chevrolet truck.	47 lb.	$2\frac{1}{2}$ tons	$\frac{1}{2}$ pint	1s.	·333	2·425	$16\frac{1}{2}$
Dept. of Interior ..	—	935	—	1254 lb.	$4\frac{1}{4}$ tons (av.)	—	—	·573	—	$27\frac{1}{2}$
F. Anderson	—	1206	1927 Chrysler tourer.	15 bags	—	$1\frac{1}{2}$ gals	2s. (av.)	·304	—	—

APPENDIXES

APPENDIX I

COMMONWEALTH OF AUSTRALIA
DEPARTMENT OF SUPPLY AND DEVELOPMENT

General Specification for Charcoal Gas Producers for Motor Vehicles

(November 1939)

1. The whole equipment shall be constructed and arranged on the vehicle as simply as possible, with particular regard to the importance of low capital cost.

2. **Guarantee.**

The manufacturer shall give a guarantee in regard to design, material and workmanship for a distance of 5000 miles, or for a period of six months (whichever occurs first), provided that the equipment is operated in accordance with the manufacturer's approved operating instructions which shall be supplied with each producer.

3. **Accessibility and Maintenance.**

The equipment shall be designed with a view to accessibility and ease of maintenance and cleaning.

4. **Safety.**

Reasonably adequate precautions shall be taken by the manufacturer to ensure safety, with special regard to fire risk and blow back of hot charcoal.

5. **Operating Range.**

The range of the vehicle without refilling of the fuel container shall be 100 miles. (The purchaser shall have the option of ordering a non-standard fuel container for an extra charge.)

6. **Starting from Cold.**

Starting from cold, the vehicle shall be capable of being driven away on producer gas alone in the minimum time, which must not exceed ten minutes. The manufacturer shall specify the method of starting and the volume of petrol, if any, required. (Special consideration will be given to those plants requiring no petrol for starting.)

7. The Weight and Volume of Apparatus.

The weigh of the whole of the equipment, excluding fuel, shall be kept to a minimum, and shall preferably be less than 15 pounds per horsepower (R.A.C. rating on petrol). As far as possible the equipment shall not be placed on the load carrying part of the vehicle unless otherwise specified by the purchaser.

8. Gas Cooler.

The temperature of the gas measured at the entry to the mixing valve shall not be more than 50° F. above the shade temperature when the vehicle is cruising at 30 m.p.h.

9. Tests.

As soon as possible, facilities will be made available for the testing of vehicles equipped with producer gas plants. The tests will be carried out by approved independent authorities for a nominal fee of £5 and certificates will be issued when the following tests have been passed:

(*a*) *Road Performance.*

The performance of the vehicle shall be measured by means of a road test of the vehicle fully equipped and loaded to its gross loaded capacity, and the performance as regards acceleration, maximum speed, manoeuvrability in traffic, and in other respects shall be reasonably satisfactory as compared with that obtained on petrol. During this road test the temperature of the gas from the gas cooler shall be measured by the independent authority carrying out the test.

(*b*) *Power Output.*

The power produced shall be measured by the independent authority on a Heenan and Froude or other dynamometer car tester capable of measuring the brake horsepower at the back wheels of the vehicle. Such tests shall be carried out at a speed corresponding to 30 m.p.h. without supercharging or change of compression ratio, and separate tests shall be made using petrol and producer gas with the same testing equipment.

During such test the gas shall be cooled to the temperature recorded on the road test at 30 m.p.h., and the power generated with producer gas shall be not less than 50 per cent of that obtained with petrol at 30 m.p.h.

(c) *Dust Separation.*

A test of four hours' duration shall be carried out by an independent authority to determine the amount of impurities passed through the cleaner, which must not be sufficient to cause excessive wear in the engine. The amount of impurities permitted shall not exceed 0·2 gramme per cubic metre of gas.

The tests will be applied to particular models of producers in relation to particular models of petrol vehicles, and with charcoal complying with the C.S.I.R. specification for wood charcoal. Such tests will be "type tests" and their certificates will be applicable to all producers of the same model in relation to the same models of petrol vehicles.

APPENDIX II

COMMONWEALTH OF AUSTRALIA
DEPARTMENT OF SUPPLY AND DEVELOPMENT

Provisional Specification for Wood Charcoal to be used in Producer Gas Vehicles

(The Commonwealth and State Governments are sponsoring the use of producer gas from charcoal as a substitute for petrol. Sir David Rivett, Chief Executive Officer of the Council for Scientific and Industrial Research, called together a Committee to draft specifications for the charcoal to be used. Subsequently, a Sub-Committee of the Conference on Problems relating to Producer Gas, Department of Supply and Development, proposed certain modifications, which were referred to the Council for Scientific and Industrial Research for acceptance. The provisional specification is now published. Copies can be obtained from the Department of Supply and Development.)

Foreword.

This specification has been somewhat hastily produced at the request of the Department of Supply and Development in order to meet an emergency created by the international situation.

It is, therefore, not to be regarded as a complete statement of the technical requirements of wood charcoal for producer gas vehicles, but rather it is to be regarded as a provisional draft, liable to alteration.

It should be understood that charcoal made in some parts of Australia will not meet all the requirements of this specification and yet will be suitable for use in gas producer vehicles. It is known, however, that operators who use charcoal conforming to this specification in their vehicles obtain satisfactory results.

Research on charcoal is now being sponsored by a number of public bodies throughout the Commonwealth, and it is hoped that a more complete specification may be produced in the near future.

To this end also, public criticism is invited. Suggestions for improvement should be forwarded to the Utilization Section, Division of Forest Products, Council for Scientific and Industrial Research, Yarra Bank Road, South Melbourne, Victoria.

1. Origin.

The charcoal shall be made from any sound wood (see Appendix A).

2. Carbonization.

The wood may be carbonized by any of the methods normally used for the production of charcoal, due regard being paid to the requirements of Clause 4 (Cleanliness) hereunder (see Appendix B).

The charcoal shall be properly and completely carbonized as indicated by its jet black colour. Charcoal containing unburnt or incompletely burnt wood, as indicated by brown pieces or patches, shall not be deemed satisfactory.

3. Appearance—Friability—Hardness.

The charcoal shall show the original texture of the wood.

It shall be firm to the touch and not easily crushed; when broken it shall not splinter or crumble, but shall present a clean fracture (for the method of testing the hardness of charcoal see Appendix C).

4. Cleanliness.

The charcoal shall be free from wood, bark and loose ash. It shall also be reasonably free from grit, dust, fines, earth and all other mineral matter so that when stirred with water as described in Appendix D solid particles separating out shall not exceed 0·5 per cent of the air-dry weight of the charcoal tested.

5. Size.

Unless otherwise specified, the standard size of charcoal to be used in producer gas vehicles shall include all material passing through a British Standard perforated plate test sieve of 1-inch aperture (side of square) and retained upon a British Standard perforated plate test sieve of $\frac{1}{2}$-inch aperture (side of square).

The amount of fines contained in charcoal for producer

gas vehicles shall be a minimum and shall be reckoned to be included in the allowable heavier than water fraction of 0·5 per cent (Clause 4).

6. Volatile Matter.

The volatile matter contained in the charcoal shall not be excessive. It is not yet known what are the permissible limits for Australian charcoals. The determination shall be carried out as in Appendix E.

7. Ash Content.

The ash content of the charcoal shall include all incombustible mineral matter present in the charcoal as received, and shall not exceed 4 per cent of the air-dry weight of the charcoal. The determination shall be carried out as in Appendix F.

Note. It is obvious that the fusion point of the ash of charcoal to be used in gas producers is of importance. In the absence of any data regarding the fusion characteristics of charcoal from Australian woods and regarding their behaviour in gas producers, however, specification of this property has been omitted. It is hoped that such data may be available for a later draft.

8. Moisture Content.

The charcoal shall be kept as dry as practicable.

The moisture content will usually be from 5 to 10 per cent of the air-dry weight of the charcoal. On account of the variation in moisture content it is desirable that if the percentage of moisture is more than 10 per cent the buyer should have the alternatives of refusing the consignment or of obtaining a rebate for the moisture content above 10 per cent. The determination shall be carried out as in Appendix G.

9. Combustion—Calorific Value.

A sample of the charcoal when lighted shall burn with a short bluish flame, free from smoke or odour.

When desired, the calorific value shall be determined as in Appendix H.

10. Sampling.

Samples shall be taken by the purchaser from the batch or consignment. Charcoal of standard size (½ inch-1 inch) and conforming approximately to the other requirements of this specification, shall be sampled by the collection of a *minimum* number of 24 portions. The portions shall be distributed evenly over the whole consignment, and the minimum weight of each portion shall be 2 lb., so that the minimum weight of the gross sample shall be 48 lb. It should be recognized that if fewer portions are taken the sampling will then be less reliable. As far as possible, each portion shall contain quantities of dust fines and smaller and larger pieces representative of the bag or section of the batch from which the portion was removed. The gross samples shall then be reduced by the method of quartering, as described in Appendix I to a sample of about 12 lb. This shall be stored in a dust-proof bin, with the exception of the sample to be used for moisture determinations, which shall be stored in an air-tight container, the samples being drawn upon when required.

Every test mentioned in the appendices shall be conducted upon a representative sample of charcoal drawn from this test sample, apart from the sample for the shatter test which shall not be obtained by quartering, but separately by careful hand picking of averaged sized pieces from the main consignment.

Note on the Reactivity of Charcoal

The reactivity of a carbonaceous fuel may be defined as the volume of carbon monoxide formed by the passage, at a known rate, of a given volume of pure carbon dioxide through the fuel when heated to a given temperature. The standard conditions used in the determination of the reactivity of coke are as follows:*

Temperature of coke	950° C.
Volume of carbon dioxide passed through the coke	100 cc.
Rate of passage of the gas ..	5 cc. per minute.

It is obvious that this property of charcoal to be used in producer gas vehicles is of the greatest importance, determining as it does the rate of gasification of the charcoal for any given set of conditions in the producer.

The reactivity of beech-wood charcoal has been found to be in the region of 180 by this standard method, but as no information has yet been obtained on this subject with charcoals from Australian woods it has been decided to omit any definite specification in this draft.

* Department of Scientific and Industrial Research, Fuel Research Technical Paper, No. 18.

However, if a charcoal is found to be unsuitable for gas production it is desirable that this property be investigated in the manner described in the technical paper mentioned in the footnote.

Appendix A

Origin.

The charcoal shall be made preferably from any sound hardwood well seasoned and free from bark, but green timber shall not be deemed unsatisfactory.

Appendix B

Carbonization.

It is preferable that a metal kiln or retort, either portable, semi-portable or non-portable, be used.

With other methods of burning it is recommended that precautions should be taken for the elimination of dirt and grit.

Appendix C

Hardness Test

It is desirable that charcoal for use in producer gas vehicles be as hard as possible, the drop test method, wherein the hardness is expressed in terms of shatter indices is here described:

For charcoal of standard size (½ inch-1 inch, Clause 5) the hardness is expressed in terms of a ½-inch and a ¼-inch shatter index. (Charcoal of finer grades may be expressed in shatter indices of lower denomination.)

Ten pounds of charcoal screened by hand so that it consists entirely of particles passing through a 1-inch British Standard perforated plate test sieve is dropped on to a concrete slab 2 feet by 2 feet from a height of 6 feet. The material having been dropped, it is shovelled back into the container and dropped again, the process being repeated four times. The charcoal is then screened by hand over a ½-inch perforated plate sieve, and over a ¼-inch wire cloth sieve, and the weights of charcoal retained on each sieve are determined to the nearest ounce. The test shall be carried out three times.

The shatter indices shall be the averages of the percentages retained on the various screens.

The following indices have been determined for a good

sample of grey box charcoal and may serve as an indication of the hardness to be expected:

$\frac{3}{4}$-$\frac{1}{4}$-*inch graded charcoal*	$\frac{1}{4}$-$\frac{3}{4}$-*inch graded charcoal*
$\frac{3}{4}$-inch shatter index 60	$\frac{1}{4}$-inch shatter index 50
$\frac{1}{2}$-inch shatter index 90	$\frac{1}{8}$-inch shatter index 80

Appendix D

Cleanliness.

The test for cleanliness shall be carried out as follows:

Approximately 3 lb. of charcoal is placed in a trough of adequate size containing water and the charcoal agitated continuously for three minutes. The contents of the trough are then allowed to stand for one minute, whereupon the charcoal floating on the surface is scooped up with a perforated metal scoop. The material which has separated out to the bottom and which remains suspended in the water after this treatment shall be regarded as the fines, dust, grit, earth and other mineral matter referred to in Clause 3; and shall be filtered on a large Buchan filter using a qualitative chemical filter paper, dried in an air oven at 105° C., weighed, and its weight expressed as a percentage of the original air-dry charcoal.

If the figure 0·5 per cent, referred to in Clause 4, be exceeded by any representative sample so tested, the whole of the batch or batches of charcoal from which the sample was taken shall be washed with water in a manner similar to the method described below:

Charcoal may be washed satisfactorily in a trough about 6 feet long, 2 feet wide and 2 feet deep, filled with water. The trough is fitted with draining boards, one at each end. The charcoal, which for preference has been previously crushed and graded is fed continuously into the water at one end and scooped up on to the draining rack at the other end, the charcoal floating on the water being gently and continuously agitated.

On warm days the charcoal will be found to dry rapidly when spread out in the sun in shallow heaps. During the winter, however, drying shelters are necessary; the charcoal must be spread out in thinner layers, and drying will take longer.

Appendix E

Determination of Volatile Matter in Charcoal.*

Two sets of procedures are detailed below. The first is a

* Taken from the Department of Scientific and Industrial Research, Forest Products Research Records, No. 29.

simplified method suitable for ordinary purposes when a high degree of accuracy is not required; the second is the standard method which is similar in principle but more carefully controlled, ensuring good reproducibility of results between different operators or laboratories. The standard method is consistent with the most recent British Standards specification for the determination of volatile matter in coke.

1. Simplified Method.

Weigh 1 gramme of charcoal, ground to pass a 72-mesh British Standard test sieve, into a cylindrical translucent silica crucible, which has a lid of the same material. The crucible and lid are as illustrated in Fig. 6 of British Standards specification No. 420, with dimensions as follows:

Crucible:	Height	37 mm.
	External diameter	25 mm.
	Internal diameter	22 mm.
Lid:	Capsule type.	
	Internal diameter (overall) ..	28 mm.
	Diameter of well (overall) ..	21 mm.
	Depth of well (overall)	4 mm.

The maximum clearance of the lid when fitted to the crucible should be 0·5 mm. The weight of the empty crucible and lid should be 12-14 grm.

Add 2-4 drops of benzene to the charcoal and fit the crucible lid, heat gently for two minutes and then at a temperature of 950° C. for five further minutes. An excessively rapid initial rate of heating causes a loss of charcoal by decrepitation and entrainment in the exuded gases. The heating may be carried out over a free gas flame, or preferably in a gas or electrically-heated muffle furnace, the rate of heating being set to provide a temperature of 950° C. by previous adjustment until a crystal of potassium chromate will just melt on the bottom of the crucible. After heating, remove the crucible and cool it rapidly on an iron slab, then in a desiccator. Finally weigh, the loss in weight being the amount of volatile matter plus the moisture in the charcoal.

2. Standard Method.

The simplified method (1) is followed, using the same crucible and lid, up to and including the addition of benzene. Then:

Place two disks of asbestos, each 25 mm. diameter and 1 mm.

thick, on the inner projections of the legs of a B.S. No. 4 gas mantle support, and support the covered crucible on these disks. Transfer to a gas—or electrically-heated muffle furnace which conforms to the following requirements: The furnace should be closed at the back, except for a flue of $\frac{7}{8}$-inch diameter which projects outwards $2\frac{1}{2}$ inches from the centre of the back plate and then rises 11 inches. The front of the furnace should have a well-fitting door, preferably hinged and of the plug type. The heat capacity of the muffle should, with an initial temperature of 950° C. permit a minimum temperature of 935° C. to be regained within three minutes from the insertion of a charged crucible and the stand. The temperature should be measured by means of a base metal thermocouple not thicker than 19 S.W.G. (1 mm.), arranged so that the bare junction is midway between the base of the crucible on the stand and the floor of the muffle. The couple and milli-voltmeter should be checked periodically. For control purposes a fixed sheathed couple may be used provided its readings are correlated at intervals with those of the standard couple.

Before the test, the furnace temperature should be adjusted to a steady value of 950° C. with an empty crucible and lid, and a support, inside, and the door closed. After this the crucible and stand should be removed, the door closed, and the furnace allowed to regain a steady temperature. Then insert the charged crucible with lid and stand, close the door and heat for exactly seven minutes, observing that at least 935° C. is regained within three minutes. After the heating period proceed as in the simplified method.

It will be observed that in the standard method the crucible is not subjected to gentle heating before introduction into the furnace; this is because the asbestos disks ensure a satisfactory rate of heat transference from the muffle to the crucible.

Note. A platinum crucible of approximately the same dimensions could be used instead of the silica crucible specified.

Appendix F

Determination of Ash.

A 2-gramme sample of air-dry charcoal is weighed into a porcelain crucible and heated (without the lid) in the oxidizing flame of a Bunsen burner. Heating is commenced gently in order to prevent spitting, due to the expulsion of the volatile materials. The crucible is heated to constant weight and the ash expressed as a percentage of the weight of the air-dry sample.

The determination is carried out in duplicate and the average percentage stated.

Appendix G

Determination of Moisture Content.

The determination is carried out on a sample of charcoal crushed to not less than $\frac{1}{8}$ inch and not more than $\frac{1}{4}$ inch as determined on British Standard $\frac{1}{8}$-inch and $\frac{1}{4}$-inch test sieves respectively. Two grammes of this sample are weighed into a weighing bottle fitted with a ground glass stopper and the bottle heated (with stopper removed) in an air oven at 105° C. to constant weight. The stopper is replaced, and the bottle cooled in a desiccator over calcium chloride and weighed. Moisture is expressed as the percentage loss in weight on the air-dry weight of the sample.

Appendix H

Determination of Calorific Value.

(Having specified ash, volatile, and moisture contents of a charcoal, the specification of calorific value is not essential. This appendix is included, nevertheless, in case it is desired at any time to carry out such a determination.)

The calorific value of the charcoal is determined using an air-dry sample ground to pass a British Standard 72-mesh test sieve.

The calorific value is determined in the bomb calorimeter in the normal manner.*

The calorific value so obtained is the gross calorific value of the charcoal under conditions of constant volume for the products of combustion; and for charcoal, in accordance with this specification, will be in the region of 11,000 B.Th.U. per 1°.

Appendix I

Note on Method of Sampling.

It has been found, in the sampling of small fuel, that for the composition of the sample to conform with the composition of the original batch to a desired degree of probability, one must remove a certain minimum number of

* See, for example, British Standards Institution's Specification No. 735, 1937, p. 42.

increments from the batch; this minimum depending upon the "average error" or degree of heterogeneity of the fuel. It should be noted that the number and weight of the increments, and hence the weight of the gross sample, bear no relationship to the total weight of the fuel sampled, but depend entirely upon this "average error" of the fuel.

The "average error" in turn depends upon the percentage of the most variable constituent of the fuel which it is desired to determine. It also depends to some extent upon particle size. It may be shown that, for the analysis of a 1-inch graded charcoal sample containing approximately 10 per cent volatile matter (the most variable constituent of the fuel) to approximate 99 times out of 100 to within 1 per cent of the true average percentage, it is necessary to select at least 24 increments of at least 2 lb. weight each from the batch.*

Reduction of Gross Sample.

(i) The whole of the gross sample shall be placed on a hard non-porous floor and mixed by heaping and turning over on to a different place two or three times.

(ii) A conical heap shall then be formed by depositing each shovelful on the top of the preceding one, so that the material slides down the cone evenly over all its surface.

A new cone shall then be formed twice in a similar way taking care to work steadily round the previous cone until it is all transferred.

(iii) The third cone shall now be flattened down uniformly with the shovel and shall be quartered by marking the heap off into quarters along two diameters intersecting at right angles.

Each pair of opposite quarters shall then be shovelled into a separate heap and one of these heaps shall be rejected.

(iv) The operation (iii) shall be repeated until about 12 lb. of charcoal remain.

Appendix J

Method of Reporting Physical Tests and Analysis of Charcoal.

The following example illustrates the form in which these results should be reported.

* See "Some General Principles of Sampling," British Standards Institution's Specification No. 403, 1930.

Cleanliness.

Heavier than water fraction—
as received .. .. 1·5 per cent
after first washing .. .. 0·02 per cent.

Shatter Test.

½-inch shatter index 70
¼-inch shatter index 91

Proximate Analysis of Charcoal.

	Air-dry Basis (per cent)	Oven-dried Basis (per cent)
Moisture	9·90	—
Volatile matter	8·12	9·00
Ash	1·98	2·20
Fixed carbon (by difference)	80·00	88·80
	100·00	100·00

INDEX

Halstead Press Pty Limited, 9-19 Nickson Street, Sydney

SAVE HARMLESS AGREEMENT

Because use of the information, instructions and materials discussed and shown in this book, document, electronic publication or other form of media is beyond our control, the purchaser or user agrees, without reservation to save Knowledge Publications Corporation, its agents, distributors, resellers, consultants, promoters, advisers or employees harmless from any and all claims, demands, actions, debts, liabilities, judgments, costs and attorney's fees arising out of, claimed on account of, or in any manner predicated upon loss of or damage to property, and injuries to or the death of any and all persons whatsoever, occurring in connection with or in any way incidental to or arising out of the purchase, sale, viewing, transfer of information and/or use of any and all property or information or in any manner caused or contributed to by the purchaser or the user or the viewer, their agents, servants, pets or employees, while in, upon, or about the sale or viewing or transfer of knowledge or use site on which the property is sold, offered for sale, or where the property or materials are used or viewed, discussed, communicated, or while going to or departing from such areas.

Laboratory work, scientific experiment, working with hydrogen, high temperatures, combustion gases as well as general chemistry with acids, bases and reactions and/or pressure vessels can be EXTREMELY DANGEROUS to use and possess or to be in the general vicinity of. To experiment with such methods and materials should be done ONLY by qualified and knowledgeable persons well versed and equipped to fabricate, handle, use and or store such materials. Inexperienced persons should first enlist the help of an experienced chemist, scientist or engineer before any activity thereof with such chemicals, methods and knowledge discussed in this media and other material distributed by KnowledgePublications Corporation or its agents. Be sure you know the laws, regulations and codes, local, county, state and federal, regarding the experimentation, construction and or use and storage of any equipment and or chemicals BEFORE you start. Safety must be practiced at all times. Users accept full responsibility and any and all liabilities associated in any way with the purchase and or use and viewing and communications of knowledge, information, methods and materials in this media.